Course 500

BASIC ENGINEERING CRAFT STUDIES

Electrical Bias

E. R. BEYNON and T. J. CUMMINS

HUTCHINSON SCIENTIFIC & TECHNICAL
LONDON

2 0 0 1 2 5 7 621.3

HUTCHINSON EDUCATIONAL LTD
178–202 Great Portland Street, London W1

London Melbourne Sydney Auckland
Wellington Johannesburg Cape Town
and agencies throughout the world

First published 1971

*This book has been set in Imprint type, printed in Great Britain
on smooth wove paper by Anchor Press, and
bound by Wm. Brendon, both of Tiptree, Essex*

ISBN 0 09 107280 8 (cased)
      0 09 107281 6 (paper)

Contents

PART B

Foreword

This book is intended to meet the electrical requirements of students following Part 1 of any of the courses in the City and Guilds of London Institute's Basic Engineering Craft Studies series.

This group of courses, better known as the 500 series, moves away from the old concept of separate courses for every craft. Instead the common material for a number of trades is used to make up one section of the course. To this is added the specialised information needed for a particular craft to make up the second section. This ensures that if later a change in specialisation is needed an adequate basis will be available from which to start any further studies which may be needed.

Part A, consisting of Chapters 1–9, contains all the material required by students who do not intend to specialise in electrical subjects, plus a few items which cannot be conveniently dealt with separately. Part B, consisting of Chapters 10–14, contains all the additional subject-matter required by students following the Electrical Bias.

Since essay-type questions for craft students are rapidly being replaced by objective testing some of the exercises at the end of each chapter are in this form.

Symbols, units and abbreviations are based on the recommendations of the Institution of Electrical Engineers and the British Standards Institute. Practical Work is based on the 14th metric edition of the IEE Regulations for the Electrical Equipment of Buildings.

E.R.B. and T.J.C.

Abbreviations and Units

List of Symbols and Abbreviations

Quantity	Symbol	Unit name	Unit symbol
Capacitance	C	farad	F
		microfarad	μF
Charge or quantity of electricity	Q	coulomb	C
		ampere-hour	Ah
Current			
Steady or rms value	I	ampere	A
Instantaneous value	i		
Maximum value	I_m		
Difference of potential			
Steady or rms value	V	volt	V
Instantaneous value	v		
Maximum value	V_m		
Electromotive force	E	volt	V
Energy	W	joule	J
		watt-hour	Wh
		kilowatt-hour	kWh
Force	F	newton	N
Frequency	f	hertz	Hz
Magnetic flux	Φ	weber	Wb
Magnetic flux density	B	tesla	T
Power	P	watt	W
		kilowatt	kW
Resistance	R	ohm	Ω
Efficiency	η		

Abbreviations for multiples and sub-multiples of units

Multiplication factor	Prefix	Symbol
1 000 000	10^6 mega	M
1000	10^3 kilo	k
0·001	10^{-3} milli	m
0·000 001	10^{-6} micro	μ

Part A

Part A

1
The safe use of electricity

Need for wiring regulations

Increasing use of electricity

So many things in our modern society make use of electricity that the demand for it increases with every year that passes. The extent of this increase is illustrated by Fig. 1.1. The diagram compares the quantity of electrical energy generated in one year during 1920 with

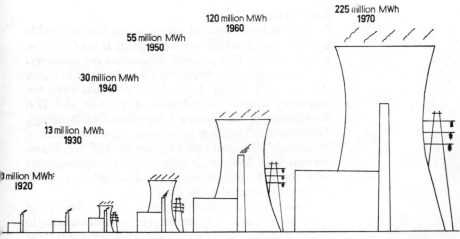

225 million MWh
1970

120 million MWh
1960

55 million MWh
1950

30 million MWh
1940

13 million MWh
1930

0 million MWh:
1920

Fig1.1 Increased consumption of electricity

that generated in the first year of every ten-year period since. Nothing known at present shows any likelihood of the growth rate slackening in the foreseeable future.

As people make use of more electrical equipment the problem of protecting them from danger becomes greater. This task is not made easier by the fact that electricity cannot be seen and must therefore be detected by the effects it produces.

Electrical hazards

Outside the electrical industry people seem to be divided into two distinct groups. One consists of those who regard electrical engineering as a vastly overrated occupation involving a few simple operations such as repairing fuses and fitting plug-tops to appliances. The second group adopts an attitude of fear and reverence bordering on superstition. The sensible approach, of course lies somewhere between the two extremes. Care combined with a reasonable knowledge of the possible risks can make electricity as safe to use as any other form of energy.

The possible risks to be considered are of two kinds, those affecting life and those affecting property. The first kind is known as electric shock. This is dealt with in detail in Chapter 2. It occurs when a body becomes part of an electric circuit. It is interesting to note that some animals, particularly cows, are much more sensitive to electric shock than are human beings.

Damage to property can be caused by the breakdown of electrical machinery or by fires resulting from overloaded, badly installed or, very rarely, faulty cables or equipment.

Regulations

In order to keep these risks down to the lowest possible level the way in which electrical energy is used must be controlled. For this purpose regulations are necessary. The regulations with the widest effect are the Electricity (Factory Act) Special Regulations 1908 and 1944, the Electricity Supply Regulations 1937 and the IEE Regulation for the Electrical Equipment of Buildings, 14th edition. The first two, being based on 'Acts of Parliament', are legally binding, but the IEE regulations are not. Such, however, is their standing that compliance with the IEE Regulations is required before a public supply is connected to an installation.

Besides those mentioned above there are other sets of regulations particularly applicable to installations in ships, churches, mines and so on. To supplement these the British Standards Institute publish 'Codes of Practice' which give advice on the best methods and materials to be used in designing and carrying out various types of installation.

Recommended supply voltages

Voltage

In Chapter 4 the exact meaning of the term 'voltage' is discussed. At this point all that is necessary is to regard it as a pressure which sets a current of electricity flowing. It is more precisely termed electro-motive force (e.m.f.) or potential difference (p.d.) and the unit in which it is measured is the volt.

Nearly all of the electricity used in industry is generated by means of rotating machinery and in theory each generator could have a different output voltage.

Public supply

In the early years of the electrical industry local authorities and private companies built their own power stations and provided electricity for their locality. Some supplied direct current (d.c.) and some alternating current (a.c.) and the voltages provided for consumers varied from place to place. This was not very convenient because equipment suitable in one part of the country could be useless elsewhere. At that time the variations were not too important, but as the number of consumers increased standardisation became inevitable.

Now as a result of the nationalisation of the electricity supply industry all major generating stations are connected into a national supply network. The stations and major transmission lines are operated by the Central Electricity Generating Board (CEGB) and the Scottish Boards. Distribution to individual consumers is handled by area boards such as the Midland Electricity Board.

'Grid' and 'Supergrid'

When electrical energy is moved from one place to another the highest possible voltage is used. This is because at high voltages smaller currents are needed to transmit given amounts of power and the losses in transmission vary as the square of the current flowing. The public supply is an alternating current system, as with alternating current, transformers can be used to obtain easily any voltage required.

Each generating station contains a number of main alternators generating at 11 000 V (volts) or some other convenient voltage. The output from the stations is transformed up and fed into an interconnecting system of overhead lines and underground cables. These enable electricity to be transmitted from the stations to places where it is needed. The greater part of the system works at 132 kV (1 kV = 1000 V) and is known as the 'Grid'. In addition there is a section known as the 'Supergrid'. Part of the 'Supergrid' has a working voltage of 275 kV but most of it operates at 400 kV.

Supplies for the area boards are taken out of the transmission system at 'Grid' sub-stations where they are transformed down to 33 kV and then 11 kV for local distribution. Many large consumers take their supply at 11 kV.

Supplies for smaller consumers

At local sub-stations the voltage is stepped down once more and in towns underground cables take the supply to small consumers. These cables have four cores (Fig. 1.2). Three of the cores are known as lines and the fourth is called the neutral. Most consumers will take a two-wire supply obtained by 'tapping off' any one line and neutral. This is called a single-phase supply (s.p.). Other consumers are 'tapped off' all the lines and the neutral. The supply in that case is called 'three-phase and neutral' (t.p. & n.). The neutral is connected to earth at the supply transformer. Voltages throughout the

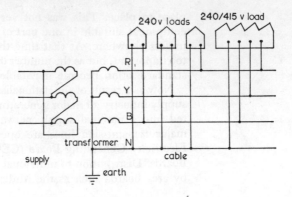

Fig1.2 Two and three phase supplies

United Kingdom are standardised for small consumers at 240 V between line and neutral and 415 V between any two lines. Single-phase supplies are used for lighting, socket outlets, heaters and similar applications. Three-phase supplies are used for electric motors.

Other voltages
Besides those mentioned in the previous paragraphs various other voltages are used. For example 4 V, 8 V or 12 V may be used for bell circuits and 110 V for temporary installations on building sites. To avoid confusion the IEE apply the following terms to the different voltages.

Extra-low voltage
Up to 50 V between conductors or up to 30 V a.c. or 50 V d.c. between conductor and earth.

Low voltage
Above extra low and up to 250 V between conductors or to earth.

Medium voltage
Above 250 V and up to 650 V.

Electrical protection

Excess current
The transfer and use of electrical energy involves a number of risks and much of the electrician's work is concerned with providing safeguards against them. The most common of these risks from the consumer's point of view is for circuits to carry more current than that for which they were designed. All equipment has a specific *current rating*; this is the current which it can carry without any ill-effect. When a piece of equipment

carries more than its rated current it is said to be carrying *excess current* and it will tend to overheat.

The methods used to provide protection against excess current are complicated by the fact that there are a number of different causes for it producing varying degrees of excess. In lighting and heating circuits it can be caused by installing larger lamps or heaters. In motor circuits excess current can result from increasing the mechanical load, also much larger currents flow when the motor starts than when it is running. Excess currents of this type are called overloads.

If for any reason the insulation between line and neutral or between two lines breaks down a very large current will flow. This is called a short circuit. Similarly large currents may flow if a breakdown occurs between line and earth. Overloads must be treated differently from short circuits and high current earth faults in providing protection.

Overload protection

Circuits can be protected against overloading by means of fuses or circuit breakers. Both are devices to interrupt an overloaded circuit without allowing any damage to the cables or equipment. The fact that a fuse or circuit breaker can protect a circuit against overloads is no indication that it will protect it against short-circuit faults.

The fuse

The cheapest form of overload protection to install and use is the fuse. Whenever current flows in a conductor its temperature rises. A large current will produce a bigger rise in temperature in a particular conductor than a small current. It will take a greater current to produce a given rise in temperature in a conductor with a large cross-sectional area than in a conductor of the same material but a small cross-sectional area. If the metal is heated enough it will melt. The principle of operation of a fuse depends on all these things. The working part of the fuse, which is called its element, is a wire of suitable size and material connected in series with the circuit it protects. When the normal working current flows it will warm up slightly. If, however, excess current flows the element will overheat and melt, breaking the circuit and stopping the flow of current. The time taken for the fuse element to melt will be shorter for a large overload than for a small one. Fuse elements are usually made of tinned copper, but sometimes of lead-tin alloy wire for smaller currents.

The fuse element is fitted into a carrier which can be

plugged into a base unit. This allows the element to be easily replaced if necessary. Fuses may be mounted in switch fuses, fuse switches, consumer control units or distribution boards (Fig. 1.3). They may also be fitted into items of equipment and plug-tops for portable appliances.

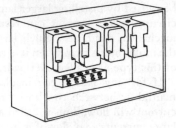

Fig1.3 Distribution board

Rewirable and cartridge fuses

Two kinds of fuse are used for overload protection. In one the fuse element, which is a bare wire, is fitted to its carrier by means of screwed terminals. This is called a *rewirable fuse*. The fuse element in the other type is enclosed in a protective case. Fuses with this arrangement are known as *cartridge fuses* (Fig. 1.4).

The advantage of a *rewirable fuse* is that when it burns out only a short length of suitable wire is needed to

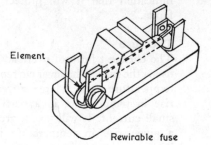

Element

Rewirable fuse

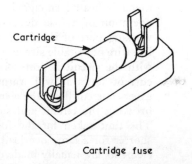

Cartridge

Cartridge fuse

Fig.1.4   Types of fuse carrier

repair it. If, however, the wrong wire is used it will either melt before the circuit is overloaded or not melt when it is overloaded. Either of these makes it useless as a protective device.

If *cartridge fuses* are used as means of protecting a circuit a supply of spare cartridges must be kept. It is, however, usually much easier to replace a cartridge than a wire element. The cartridge used for replacement must, of course, have the right rating, otherwise its protective value will be lost. This is particularly important with fused plug-tops where the 3 A and the 13 A cartridges are readily interchangeable. Another important point with the cartridge fuse is that the element being totally enclosed cannot damage its carrier when it melts except in the case of short-circuit or high-current earth faults.

A disadvantage of fuses in general is that there is always some delay in repairing them and this makes them unsuitable for use where overloads are frequent.

Circuit breakers

In many installations the time taken to repair a fuse may be too great for safe and efficient operation. In such cases circuit breakers are used. These are switches which automatically open, disconnecting the supply when a fixed value of overload is reached or maintained for a given period of time.

When an electric current flows it sets up a magnetic field and also causes a temperature rise in the conductor carrying it. Either of these effects may be used to operate a circuit breaker.

A simplified magnetically operated circuit breaker is illustrated in Fig. 1.5. Its operation is as follows: the main contacts are closed against the action of the trip spring by pressing button (F). The contacts are locked in the closed position by the trip catch (C) which engages on the lower end of the plunger (D). This is free to move through the centre of the coil (E) but is normally kept in the position shown by its own weight. When the circuit is in use all the current it carries passes through the control coil (E). This current causes a magnetic field to be set up concentrated in the centre of the coil. The magnetic field tends to raise the plunger, but if the current is not greater than the setting of the breaker the plunger will not move.

When an overload occurs the magnetic field in the coil becomes strong enough to lift the plunger. Once the plunger pulls clear of the trip catch (C) the trip spring is free to expand and the mains contacts are rapidly opened, breaking the protected circuit. Once the circuit has been broken the magnetic field in the coil

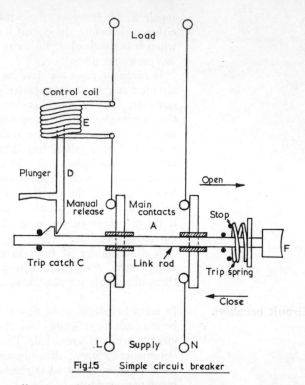

Load

Control coil

E

Plunger D

Open

Manual release

Main contacts

A

Stop

F

Trip catch C

Link rod

Trip spring

Close

L○ Supply ○N

Fig.1.5   Simple circuit breaker

collapses and the plunger falls again. The circuit breaker can then be re-set by pushing the button (F) until the trip catch re-engages with the plunger.

Additional requirements

The design of the circuit breaker shown is simplified in order to convey the basic operating principle. A practical circuit breaker would need to contain a number of extra provisions. The first of these would be some means of adjustment so that the current at which the circuit breaker opens can be altered if necessary. It would also require a delaying device by which the breaker could be prevented from tripping on overloads that did not last very long. A circuit controlling machinery will also contain a release mechanism to open it if the supply fails. The first two additional requirements are often provided for by means of a dashpot and plunger arrangement to slow down the movement of the plunger. The last requirement may be met by introducing a second coil connected in parallel with the supply and arranged so that it releases the trip when the supply is cut off.

Close and coarse protection

If the overload protection device for a circuit is such that a current of one and a half times the normal full-load current will cause it to operate within four hours, the circuit is said to have close excess-current protection. If the device requires a larger current or a longer time to operate the protection is described as coarse. In general close protection is provided by circuit breakers and coarse protection by fuses. The type of protection used affects the current rating of the cables used in a circuit.

Short-circuit protection

When a short circuit occurs it is possible for the excess current flowing to become several hundred times greater than the normal load current. No ordinary fuse or circuit breaker could interrupt such a current without an explosive release of energy which could cause considerable damage. Special fuses are used to deal with such conditions. They are known as *high rupturing capacity* (HRC) or *high breaking capacity* (HBC) fuses.

The HRC fuse consists of a number of parallel elements sealed in a tough ceramic cartridge (Fig. 1.6). When the

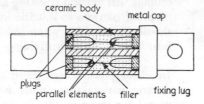

Fig1.6 HRC Fuse

elements melt, the filler around them fuses into insulating blocks which prevent any arcing. They are so designed that they do not react to a normal overload of two or three times the rated current. If, however, a fault current of the size associated with a short circuit starts to build up the fuse blows in a few thousandths of a second. This prevents the current reaching a dangerously high level. Some circuit breakers are capable of clearing short-circuit faults, but the smaller sizes are generally only useful for overload protection.

Fire risk

Fires having an electrical origin are not very frequent and are more often than not the result of misusing electrical equipment. The chance of an electrical fire can be reduced to negligible proportions by compliance with the IEE Regulations at the installation stage and common sense in the use of electrical apparatus.

The fact that conductors warm up when carrying current has already been mentioned in this chapter. Over-loading will cause overheating and this is an obvious source of fire risk which must be avoided. Badly made joints and loose connections also tend to overheat and can be dangerous. It is obvious that the heat generated in equipment must be allowed to escape if dangerously high temperatures are not to be reached. This makes it necessary to provide adequate ventilation at all times. Particular care is needed in this respect when providing extra protection against mechanical damage.

Where motors or generators are used sparking often occurs at the brushes. Sparks are also caused whenever a switch is opened or contacts are separated. These sparks are for the most part quite harmless. If, however, sparking occurs where there are explosive gases such as petrol vapour, sawdust or similar inflammable material present they could be dangerous. In such cases special machines and wiring methods must be used.

Apart from preventing fires of electrical origin it is also necessary to ensure that an electrical installation does not assist in the spread of any fire whatever its source may be. This is best done by ensuring that no holes are left unsealed where cables pass through wall, floors or ceilings. All vertical trunking and ducts should contain fireproof barriers at suitable intervals. Similar barriers should be provided where cables running under floorboards pass beneath dividing walls.

Emergency control of machines

In every place where there are machines driven by electric motors there must be a switch conveniently placed for stopping the machinery in order to comply with the Factories Act. The IEE regulations also require that every motor must be fitted with a device to prevent it automatically restarting after a supply failure if doing so could be dangerous. It must also be provided with an isolator to allow work on the machinery without danger. If the motor is rated at 375 W ($\frac{1}{2}$ hp) or more it must also have a starting arrangement including suitable excess current protection.

The isolators and emergency stop buttons in machine shops and other industrial premises should be placed where they can be seen without any difficulty. They should be painted bright red and clearly labelled to indicate what they control.

Summary of regulations

Excess-current protection: IEE Regulations Part 2 A6, A7, A10–13, A65, A67–68, C26, J4.

Fuses: IEE Regulations Part 2 A3, A6–13, A30, A32, A39, A42, A45, A48, A54, A67, A68, B123, C2, C43, C50, D22, F1, F6, F7, F11.

Circuit breakers: IEE Regulations Part 2 A3, A6–10, A14, A63–69, B123, C2, D21–26, E5, F1, G14, H1.

Fire: IEE Regulations Part 2 B35, B39, B40, B71, C8–10, C15, F1, H1, K23.

Control gear for machines: IEE Regulations Part 2
A63–65, F12.

Exercises

1 State whether the following statements are correct or not:
  a The quantity of electrical energy consumed per year in Britain has remained roughly the same for the past twenty years.   True/False
  b The 'Supergrid' transmits electrical energy at 132 000 V.   True/False
  c The public supply in England and Wales is standardised at 415/240 V a.c.   True/False
  d The term 'medium voltage' is applied to voltages between 50 V and 250 V a.c.   True/False
  e The recommended voltage for temporary installations on building sites is 110 V a.c.   True/False

2 Complete the following statements using one of the alternatives given:
  a The current carried by a piece of equipment under normal working conditions is known as its
    (i) power (ii) current rating (iii) fusing current
  b The least expensive form of overcurrent protection is provided by
    (i) the circuit breaker (ii) the HRC fuse (iii) the rewirable fuse
  c The HRC fuse provides the best protection against
    (i) short circuits (ii) overloads (iii) fire
  d Dashpots in circuit breakers prevent them
    (i) sticking in (ii) tripping too easily (iii) overheating
  e Starters must be provided for all motors having a rating greater than
    (i) 1500 W (ii) 800 W (iii) 375 W

3 Discuss the reasons which make it necessary to have regulations for controlling the use of electrical energy.
4 What are the differences between the objects of the IEE Regulations and those of the BS Codes of Practice?
5 What is the 'Grid' system? Why is it necessary and why does it make use of such high voltages?
6 Discuss the various factors that can give rise to the flow of excess current in an installation.
7 What are the relative merits of rewirable fuses and circuit breakers as a means of protection against overloads?
8 What is the purpose of an HRC fuse and how does this affect its construction?
9 What factors contribute to the risk of fire in an installation?
10 Summarise the requirements of the IEE Regulations with respect to the control of motors rated at more than 375 W.

2
The prevention and treatment of shock

Causes of electric shock

When an electric current flows through a person's body, if the current is large enough, the person will experience an electric shock. Electric shock is both unpleasant and dangerous. For a current to flow, a complete circuit is needed and anyone receiving a shock must be forming part of a complete circuit. This can be done in a number of different ways, some of which we will now examine.

Touching 'live' conductors

The current to and from a lamp, motor or other electrical appliances flows through current-carrying conductors. There will be a difference in voltage between any two of these and anyone touching two at the same time will complete the circuit and receive a shock. In Fig. 2.1

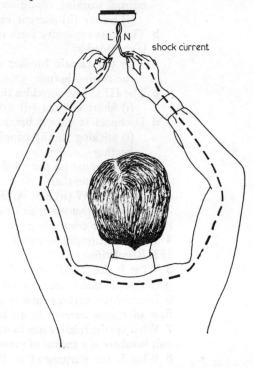

Fig 2.1 Shock from two live conductors

the electrician is touching the two live cores of a lamp flex. There is a voltage of 240 V between them and this causes current to flow through his arms and across his chest as indicated by the dotted line.

A shock caused by touching two live conductors is more likely to happen to an electrician than to anyone else. The reason for this is that all electrical equipment is designed so that live parts cannot be easily touched. This protects the normal user, but an electrician will often need to work on the current-carrying parts of equipment. For this reason an electrician must exercise a great deal of care to avoid working on equipment without disconnecting the supply.

Touching one live conductor and earthed metal
The main mass of the earth behaves in some ways as a conductor. All public supplies in Great Britain are connected to earth. It is therefore possible to have a complete circuit from the earthed point of the supply system through the supply conductors and back through any earthed metal in the installation as long as there is a link between line and earth. In Fig. 2.2 the link is

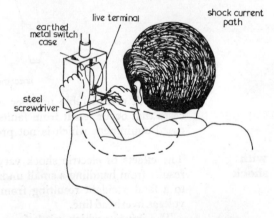

Fig 2.2 Shock between live conductor and earth

provided once more by the electrician who is touching the earthed metal framework of the switch and a live conductor at the same time. Once again the path of the current is indicated by a dotted line and, of course, anyone in such a position will receive an electric shock.

Touching 'live' unearthed metal and earth
If metalwork connected to earth comes in contact with a 'live' conductor a large current tends to flow. This will cause the fuse protecting the circuit to blow or the circuit breaker to 'trip', disconnecting the supply. If, however, a live conductor makes contact with non-earthed metal no current will flow until a link with earth is established. This is often provided by a human

body. In Fig. 2.3 the electrician is once more receiving a shock by completing the circuit between faulty non-earthed metal and the earthed metalwork of the lamp

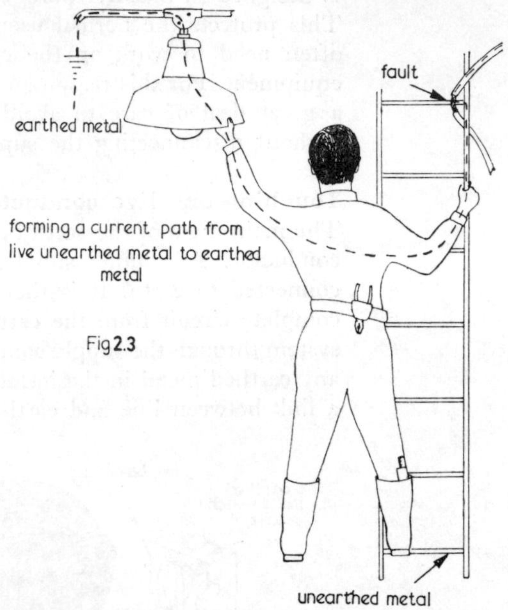

earthed metal

fault

forming a current path from live unearthed metal to earthed metal

Fig 2.3

unearthed metal

fitting. A considerable number of the domestic shock cases that occur result from faults developing in metal-cased equipment which is not properly earthed.

Dealing with electric shock

The effects of electric shock vary from the tingle that results from handling a small undischarged capacitor up to a fatal accident resulting from contact with a high-voltage overhead line.

The injuries which result from shock can be roughly divided into three categories. The first group are physical injuries such as broken bones and bruises due to the instinctive desire of the person receiving the shock to get away from the cause of their discomfort. These are particularly liable to be suffered by people working on ladders or scaffolding at the time they receive a shock and provide another reason for care in providing safety belts and guard rails.

The second type of injury is due to interruption of the working of heart and lungs which result when current passes through the nerve and muscle systems which control them. Lastly severe injury is often caused by burning in electrical accidents if the currents involved are very large. With accidents of the first and the last type the risk of lasting disfigurement or disability is quite high. With the second type death is much more likely to result, but if the victim survives he will often suffer no long-term ill-effects. Frequently cases of shock in this category will not be breathing. In such a case immediate action is necessary, as it can often restore a shock victim who is apparently dead.

Time is precious when dealing with a person who is unconscious as a result of receiving an electric shock. It is essential, therefore, that if you discover such a person you should start the revival process at once. Do not waste time looking for someone better able to deal with the emergency, as this could cost the victim his life.

It often happens that the victim falls clear of the live conductors on losing consciousness, but this is by no means always the case. Before touching the affected person, therefore, the rescuer must check that it is safe to do so. If there is still contact with live metal and the switch controlling it is not very close the victim must be pulled off. This can be done by dragging him clear by any loose clothing or levering him off with a piece of dry wood such as a broom handle. On no account should the bare skin or wet clothing be touched as this could result in the rescuer receiving a shock. Once clear of the live metal, anything tight, such as a necktie or belt, should be quickly slackened.

The victim's mouth should be checked to ensure that his tongue is not blocking the flow of air to the throat. False teeth, if any, should be removed. Artificial respiration should then be started.

Methods of artificial respiration

The best-known methods of artificial respiration are Schafer's method, the Holger-Nielson method, the Eve-Riley rocking stretcher and the mouth-to-mouth or 'kiss of life' method. Others include the use of power-assisted resuscitators which feed oxygen to the victim as well as reviving the lung action. The fact that there are so many ways of dealing with what is a comparatively rare form of accident is due to the fact that the same techniques are used for restoring persons who are apparently dead as a result of drowning. Each method has advantages and disadvantages. The one that can be most easily applied and has been used with a great deal of success is the mouth-to-mouth method.

Applying the 'kiss of life'

The victim is laid on his back and the rescuer sits or kneels beside his head. The head is held with one hand below the chin and the other at the crown (Fig. 2.4). The top of the head is pushed back and the jaw upwards and forwards. This is done to allow a clear path from the

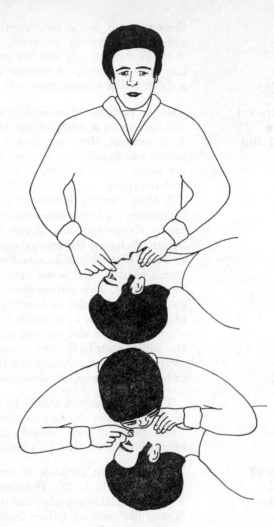

Fig 2.4 The kiss of life

victim's mouth to his lungs. One hand is kept under the chin and the other is used to close the victim's nostrils. The rescuer then takes a deep breath, seals his lips around the victim's mouth and blows steadily into his lungs until the victim's chest starts to rise. The rescuer then removes his mouth, turns his head to one side and takes another deep breath. While this is happening the shock victim will automatically breathe out. At the start of the revival process six breaths should be given as quickly as possible and then should be given every six seconds until the victim can breathe without help. If it is not possible to use the victim's mouth it should be held shut and the rescuer should breathe into the victim's nose.

After-care

When the affected person regains consciousness he should be kept warm and quiet. He may be given something warm to drink such as sweet tea, but no alcohol or other stimulants, as they would be harmful in the circumstances. Further care must be placed in the hands of a doctor.

Whether the 'kiss of life' or some other method is used it should be kept up until a doctor arrives at the scene of the accident. People have been revived after several hours of unconsciousness, so efforts should not be abandoned until a doctor takes over.

Summary of regulations

Electricity (Factories Act) Special Regulations 1908 and 1944.

Royal Society for the Prevention of Accidents. Pamphlets and posters.

Electrical Times. Shock-treatment display cards.

Exercises

1 Complete the following sentences using one of the alternatives given.

    a For current to flow a circuit must be

        (i) insulated (ii) complete (iii) protected

    b Electric shock occurs when a person links two points at

        (i) different voltages (ii) a high voltage (iii) at different temperatures

    c If a 'live' conductor connected to a public supply touches earthed metal

        (i) no current will flow (ii) current will flow to earth (iii) the supply voltage will increase

    d It is not possible to receive an electric shock as a result of touching

        (i) two lampholder terminals (ii) opposite ends of an electric fire element (iii) two well-earthed pieces of metal

    e Electric shock is

        (i) never fatal (ii) always fatal (iii) sometimes fatal

2 State whether the following are correct or not.

    a On finding someone suffering from shock the rescuer should first try to get a doctor.

                                  True/False

    b A shock victim may be revived even if apparently dead.   True/False

    c The tongue of a shock victim could restrict the flow of air to his lungs.   True/False

    d The 'kiss of life' method is the simplest form of artificial respiration.   True/False

    e After a shock victim has recovered consciousness he should be given a little brandy.   True/False

3 Describe with the aid of diagrams two different ways in which a person in the home could receive an electric shock.

4 Why is an electrician more liable to receive a shock from contact with two live conductors than any other member of the general public?

5 State some simple precautions which would reduce the risk of electric shock for people working with electrical equipment.

6 Describe briefly the types of injury that could result from electric shock.

7 Describe three different ways of removing an unconscious person from contact with live conductors without risk to the rescuer.

8 Describe the action which should be taken on finding a person suffering from electric shock.

9 What are the main advantages of the 'kiss of life' method of artificial respiration?

10 When a shock victim regains consciousness how should he be treated?

3
Electrical properties of materials

Before discussing the electrical properties of materials the simple atomic theory of matter is described, since it forms the basis for the explanations of various electrical, mechanical, physical and chemical phenomena.

Simple atomic theory

The Rutherford-Bohr theory on the atomic structure of matter is the theory mainly adopted by engineers to explain various electrical, mechanical, physical and chemical phenomena. It proposes that the substance called electricity is a fundamental part of all matter.

Elements, atoms, molecules, compounds

All matter can be subdivided into substances that have different chemical properties. These substances are known as *elements*. The smallest particle into which an element can be divided and still retain its chemical properties is called an *atom*. Atoms of a given element are all alike, but differ from the atoms of other elements.

Molecules

A molecule is the smallest particle of a substance that can exist in the free state. It may consist of

1 a single atom of an element
2 two or more atoms of the same element, or
3 two or more different atoms combined to form a molecule of a *chemical compound*

Electrons, protons, neutrons, ions

In the Rutherford-Bohr theory an atom is regarded as a miniature solar system in which a central nucleus is enveloped by minute particles in orbit called electrons. An electron has a definite mass and carries a fixed negative charge of electricity. Electrons are similar in any atom, no matter to what element the atoms belong.

The nucleus of an atom contains particles known as *protons* and *neutrons*. It also contains other particles which are ignored in the simple atomic theory under discussion. A proton is an electrically charged particle of mass approximately 2000 times that of an electron. It carries the same amount of electrical charge as an electron but the charge is of the opposite kind, being positive. A neutron has a mass equal to that of a proton, but no electrical charge.

When an atom is in its normal state—that is, electrically neutral—the number of electrons equals the number of protons. The total negative electrical charge of the orbital electrons is neutralised by an equal amount of positive charge of the protons in the nucleus. If an atom loses an electron, it is known as a *positive ion* since it carries a surplus positive charge of electricity. Similarly when at atom gains an electron, it is called a *negative ion* since it has a surplus negative charge. The process of forming ions is known as the *ionisation process*. Ions are identified by the chemical name of the element but they do not exhibit its chemical properties.

Since protons and electrons are present in all matter and carry opposite kinds of electricity (positive and negative electrical charges), the atomic theory claims that electricity is a fundamental part of all matter. The mass of an electron is so small compared with the mass of a proton that it may be neglected and the electron regarded as a particle of electricity. Thus the electron is the ultimate particle of electricity. It is the naturally occurring unit of electricity that exists in all matter. For most practical purposes, the electric charge of an electron is too small an amount in which to measure a quantity of electricity. To attempt to do so would be equivalent to measuring a quantity of sand by counting the number of grains, rather than by weighing it.

The unit of quantity of electricity or charge generally employed is the *coulomb*. A coulomb is the total charge carried by 6.3×10^{18} electrons or protons. Thus, the magnitude of the charge carried by a single electron or proton is 1.6×10^{-19} coulomb.

Law of electric charges

A fundamental property of electricity is that unlike electric charges (i.e. positive and negative) attract each other, while like charges repel each other (see Fig. 3.1).

Fig.3.1 Law of electric charges

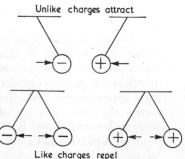

Unlike charges attract

Like charges repel

Fig 3.2 represents the general structure of atoms of all elements. The regions in which the electrons travel round the nucleus are called *shells*. The nucleus is a compact mass of neutrons and protons enveloped by clouds or shells of electrons. These shells are lettered from the nucleus outwards, K, L, M, N, O, P, and the

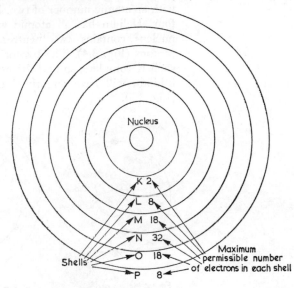

Fig.3.2 General structure of atoms

numbers shown in the diagram indicate the maximum permissible number of electrons for each shell. The outermost shell of any atom has a maximum of eight electrons. For shell M to contain eighteen electrons, there must be at least one electron in shell N. If shell N is the outermost shell of an atom, then the maximum permissible number of electrons in shell N is eight and not thirty-two. The *atomic number* of an element is the number of protons or the number of electrons in an atom of that element. The *atomic weight* of an element is the total number of neutrons and protons in the nucleus of an atom of that element.

The simplest atom is one of *hydrogen* (Fig. 3.3).

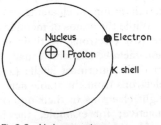

Fig 3.3 Hydrogen atom

Hydrogen has an atomic number of one and an atomic weight of one, since its nucleus consists of only one proton. It is the first in the table of elements (the Periodic Table) where all known elements are listed in order of their atomic number.

The rare gas *helium* comes next in the periodic table with an atomic number of two, but an atomic weight of four. Helium has an atomic weight of four since its nucleus contains two neutrons in addition to two protons (Fig. 3.4). The two orbital electrons make the shell K complete, so that the next element in the table, lithium, will have an orbital electron in the shell L (Fig. 3.5).

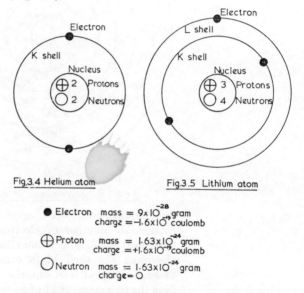

Fig.3.4 Helium atom Fig.3.5 Lithium atom

● Electron mass = 9×10^{-28} gram
charge = -1.6×10^{-19} coulomb

⊕ Proton mass = 1.63×10^{-24} gram
charge = $+1.6 \times 10^{-19}$ coulomb

○ Neutron mass = 1.63×10^{-24} gram
charge = 0

Lithium has an atomic number of three and an atomic weight of seven and is the lightest metallic element. As the atomic number increases by one, an additional proton occurs in the nucleus and an additional electron in orbit.

This process may be continued until all the elements in the table have been covered. By studying the atomic structural diagrams and the physical and chemical properties of the elements similarities are revealed in certain groups of elements.

While the atoms of metallic elements usually have one or two electrons in their outermost shells, those of non-metallic elements usually have six or seven electrons in their outermost shells (a deficiency of one or two electrons from the stable octet). The elements that have eight electrons in their outermost shells are chemically inactive; for example, the inert gases, neon, argon, and krypton.

Chemical reactions in which different elements combine to form chemical compounds involve the sharing of electrons in the outermost shells of the atoms of the elements. The molecules of the compounds formed in this manner have stable outermost shells containing eight electrons. These outermost electrons are known as *valency electrons*. In the metallic elements the valency electrons are easily dislodged from the parent atom because they are screened from the attraction of the positive nucleus by the intervening negatively charged shells or clouds of electrons.

If an element is chemically active, the atoms usually combine to form molecules of the element by sharing the outer electrons. The ability to share outermost electrons is known as valency. The atom of sodium has one outermost electron which it readily shares to form a molecule of another compound, so it is said to have a valency of 1, or to be *univalent*. The atom of chlorine also needs to share one electron to form a stable octet of outermost electrons, so it also is univalent. Thus an atom of sodium and an atom of chlorine share outermost electrons and combine to form a molecule of sodium chloride or common salt.

Solids, liquids and gases

There are three states of matter: the solid, the liquid and the gaseous. A substance is known as a solid, a liquid or a gas according to its physical form at ordinary temperatures.

The physical state of a substance is a guide to the manner in which the atoms of molecules are packed together in a given volume at normal temperatures, and to the freedom of movement of atoms within that volume.

In solids the atoms or molecules are packed closely together, and they hold fixed positions in relation to each other. The forces of attraction between the atoms or molecules of a solid are very large, and these cohesive forces resist any change of shape in the solid. For some substances the geometrical arrangement of the atoms fixed by the inter-atomic forces produces a crystalline structure. This fixed pattern is referred to as the crystal lattice of the solid. The metallic elements possess crystalline structures, and the cubic crystal lattice of copper (Fig. 3.6) is known as a *faced-centre cube*.

Models of crystal lattices may be made in which wooden spheres represent the atoms and a framework of springy wire simulates the cohesive forces or bonds between them. At normal temperatures the atoms vibrate about the fixed positions determined by the geometrical pattern of the crystal lattice. This behaviour may be demonstra-

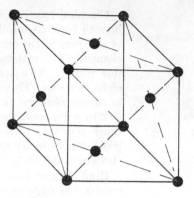

Fig.3.6 The crystal lattice of copper

ted by giving a model of the crystal lattice a light blow to set the spheres vibrating on the wire framework. The speed of the atomic vibrations depends on the temperature of the substance, as the speed of the model's vibrations depends on the intensity of the blow given to it.

Solids

If sufficient heat energy is supplied to a solid, the temperature increases and the atoms vibrate more vigorously so that the solid expands, melts and takes a liquid form. Part of the heat energy has been absorbed in moving the atoms further apart; in other words in expanding the solid, and as the cohesive forces are thus weakened, a softer material results. Another part of the heat energy increases and the speed of vibration of the atoms. This heat energy is stored as the kinetic energy of the atoms, that is, the energy the atoms possess by virtue of their motion. As more heat energy is supplied the lattice bonds are stretched further, and the kinetic energy increases until the bonds eventually break. When this happens the crystal lattice collapses and the atoms move at random in the material. Thus, the solid changes into its liquid form. As the transition from solid to liquid takes place, some heat energy is supplied without raising the temperature since it is used to break the lattice bonds. This amount of heat energy is known as the latent heat of fusion.

Liquids

In liquids the atoms or molecules move at random and the cohesive forces are small, so that there is little opposition to change of shape. Again, the speed of the atomic or molecular movements depends on the temperature of the liquid. An atom or molecule near the surface of the liquid experiences a force of attraction due to the other atoms in the liquid. This force prevents the atoms leaving the liquid by pulling them back from the surface towards the centre of the liquid. This effect is known as the surface tension of the liquid.

If more heat energy is supplied, the liquid absorbs the energy by increasing the speed of the atoms and the distance between them. This is shown by a rise of temperature and the expansion of the liquid, which changes into its gaseous form when the atoms acquire sufficient kinetic energy to weaken the bonds between the atoms yet further, and thus to escape the boundaries of the liquid to form a gas. Some heat energy is required to break down the cohesive forces at this point so that when the liquid changes to its gaseous form at boiling point, heat energy is expended with no temperature rise. This heat energy is known as the *latent heat of vaporisation*.

Gases

In gases the atoms are further apart than in liquids or solids with a corresponding reduction in the interatomic forces of attraction and the resistance to change of shape. The atoms or molecules of a gas move at random with a greater average speed than in the liquid state. A pressure is exerted on the container by the gas because of the collisions of the random moving atoms or molecules with the sides. The temperature of the gas determines the average speed of its atoms or molecules. If the temperature is increased by supplying the gas with more heat energy, the average speed of the atoms or molecules increases. The additional energy is stored as the increased kinetic energy of the atoms or molecules of the gas. As the temperature increases so the pressure of the gas increases, because the rate of collisions between the atoms or molecules and the container goes up.

To solidify liquids, and to liquefy gases, heat energy must be extracted from them. Heat energy is taken from a substance at the expense of the molecular, kinetic energy resulting in the restriction of movement of the atoms or molecules. The liquids and solids then re-form from the gaseous and liquid states.

The changes of state may be observed when ice melts to form water, which if boiled changes to steam. The steam may be condensed to form water, which solidifies when frozen to give ice.

Conductors and insulators

A material is classified as either a conductor or a non-conductor (an insulator in the electrical sense) depending on whether it will or will not conduct electricity.

To conduct electricity, the material must allow electrical charges to pass through it. The amount of electric charge that crosses a section of the conducting path in one second is known as the *electric current* through the material. Thus, an electric current is the rate of flow of electric charge. The unit of current is the ampere, and one ampere passes through the material when one coulomb/second is the rate of flow of electric charge in the material.

A good conductor is a material that allows an electric current to pass easily through it. A perfect insulator is a material that will not allow an electric current to flow through it. These are the two extremes, but there are other materials known as *resistors* through which current passes with difficulty, and some other materials called *semi-conductors* which under certain conditions allow a limited current to pass through them.

The reason why some materials are conductors and some are insulators may be found from a study of their atomic structures.

In the outermost shells of their atoms, good conductors such as the metallic elements copper, silver and aluminium have electrons that are only feebly attracted by the positive nuclei. This is because they are screened by the intervening shells or clouds of electrons. These outer electrons are free to move at random in the spaces of the crystalline lattice of the solid conductor, and are sometimes referred to as the *electron gas* in the material (see Fig. 3.7a).

The atoms having lost their outer electrons become positively charged ions. If a definite directional drift is imposed on the random movement of the free electrons, then an electric current will flow through the material. The drift may be created by connecting the ends of the conductor to the positive and negative terminals of a cell or battery (see Fig. 3.7b). Since unlike electric charges attract, the negative charges of electricity (the electrons) will drift towards the end of the conductor connected to the positive of the cell. If the positive ions were free to move, they would tend to migrate towards the negative battery terminal. This would mean bodily movement of the conductor but the relative positions of the ions are fixed in the crystal lattice and therefore no bodily movement of the conductor is possible. Thus an electric current through a solid conductor is a flow of electrons from the negative to the positive end.

The atoms or molecules of insulator materials have very few free electrons. In other words, there are hardly any free electrons available to drift through the material under the influence of the battery if the insulator material is connected to it.

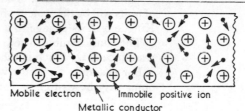

Mobile electron   Immobile positive ion

Metallic conductor

Fig.3.7(b) Metallic conductor carrying an electric current

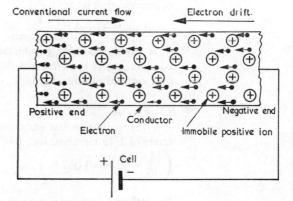

Conventional current flow   Electron drift.

Positive end   Negative end

Electron   Conductor   Immobile positive ion

+ | Cell
  | —

The property of electrical conductivity depends on
the degree of ionisation caused spontaneously as the
outer electrons break free from their atoms and wander
at random in the crystal lattice of the material. Good
conductors have a high degree of spontaneous ionisation,
but in poor conductors or resistors it is moderate, and
in insulators very small.

Conductors, insulators and resistors make it possible
to form insulated electric circuits. Then the currents
are confined to the conductors by the insulators, and are
controlled by the resistors.

**Electrical
properties of
materials**

The electrical properties discussed at this stage are
resistance, conductance, resistivity, conductivity, and
insulation resistance.

Resistance (symbol: R, r; unit: ohm)
The opposition offered by a material to the axial drift
of electrons through it—that is, to an electric current
passing through the material—is known as its electrical
resistance. This is the property of the material that converts electrical energy taken from the current into
thermal energy of heat.

The symbol for resistance is R or r. In the system of
units named 'Système International d'Unités' and for

41

which the international abbreviation is 'SI' the unit of
resistance is the ohm. The International System (SI)
of units will be the system of units adopted by the United
Kingdom in its change to a metric system of units.

The unit symbol used after numerical values of
resistance is Ω, which is the Greek capital letter omega.

For example, a resistor which has a resistance of 15
ohms may be written as 15Ω.

Conductance (symbol: G; unit: siemens)
With direct currents conductance is the reciprocal of
resistance, that is:

$$\text{Conductance} = \frac{1}{\text{resistance}}$$

G is the symbol for conductance and the SI unit of
conductance is the siemens. After numerical values the
unit symbol for siemens is S

$$\therefore \quad G = \frac{1}{R} \text{ siemens}$$

The resistor that has a resistance of 15Ω has a conductance of 1/15 siemens, i.e., 0·067 S.

$$\left(\frac{1}{15} = \frac{1}{15} \text{ S} = 0·067 \text{ S} \right).$$

Resistivity (symbol: ρ; unit: ohm metre)
Resistivity of a material or more precisely its volume
resistivity is the resistance between opposite faces of a
unit cube of the material at a given temperature (see
Fig. 3.8). Its symbol is ρ, the small Greek letter rho.
In SI the unit of resistance is the ohm and the unit cube
is a metre cube. Consequently the SI unit of resistivity
is the ohm metre. After numerical values of resistivity
the abbreviation or unit symbol used for ohm metre
is Ωm.

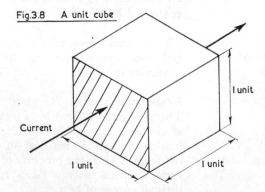

Fig.3.8   A unit cube

1 unit

Current

1 unit

1 unit

Volume resistivity or resistivity is sometimes referred
to as the specific resistance of a material.

Conductivity (symbol: γ; unit: siemens/metre)

Conductivity is the reciprocal of resistivity. The symbol for conductivity is γ, the small Greek letter gamma.

$$\text{Conductivity} = \frac{1}{\text{resistivity}}$$

In symbols $\qquad \gamma = \dfrac{1}{\rho}$

The SI unit of conductivity is the siemens/metre which is abbreviated after numerical values by the unit symbol S/m.

(Units of $\gamma = \dfrac{1}{\Omega\text{m}} = \text{S/m}$)

Insulation resistance (or insulance)

Insulation resistance or insulance is the resistance under prescribed conditions between two conductors or systems of conductors normally separated by an insulating material. It is in effect the resistance of the insulation under the prescribed conditions. A good insulator is a material with poor conductivity and low conductance. It has a high resistivity and possesses high resistance. Good insulators such as fused quartz, mica, rubber and mineral oil have resistivities of the order of 10^{12} ohm metre and insulation resistances of millions of ohms, that is, of megohms. These high insulation resistance values keep the leakage currents through the insulating materials to very low values. A material which is a poor conductor and therefore a good insulator is also called a dielectric.

Other properties of materials

Before a material is used for a particular electrical application remember that its physical, mechanical and chemical properties are also important and must also be considered with its electrical properties.

The physical properties usually considered are the normal state of the material, that is, if it is a gas, a liquid or a solid at normal temperatures, its thermal stability, that is, its behaviour at various temperatures, also for insulating materials or dielectrics their ability or inability to absorb moisture is also important since its resistance falls with moisture absorption.

For solid materials the mechanical properties to be considered are their tensile, compressive, shearing and for thin sheets their tearing strengths. Also important are the elasticity, ductility, hardness and toughness of the materials. Elasticity is the ability to resist permanent changes of shape. Ductility is the material's ability to be drawn into wires of small cross-section by a tensile force. Hardness is the property that resists wear. Toughness is the quality that resists fracture due to suddenly applied loads, that is, to impact loads.

Chemical properties of materials found together in a particular electrical application must be such that no adverse chemical reaction takes place between those materials.

Exercises

1 Define **a** an element
   b an atom
   c a molecule
 and **d** a compound

2 What is **a** an electron
    b a proton
    c a neutron
 and   **d** an ion?

3 State the law of electric charges.

4 Why are some substances either solid, liquid or gaseous at normal temperatures according to the simple atomic theory?

5 Give an explanation using the simple atomic theory for the changes in state that occur in a solid substance when it is heated.

6 Explain with reference to their atomic structures why some materials are electrical conductors whilst others are insulators.

7 What is the essential difference between conductors and insulators? Under headings of conductors and insulators classify the following materials: lead, carbon, paper, bakelite, mercury, porcelain, asbestos, mica, tungsten. Give a practical example of the use in electrical work of *one* conductor and *one* insulator from the above list.

8 Define the following properties of materials:
  a resistivity
  b conductivity
 and **c** insulation resistance

4

Basic electrical theory

**The electron
theory**

The simple atomic theory discussed in Chapter 3 shows that electricity is present in all matter as negatively charged particles called electrons that are in orbit around positively charged nucleii. Electricity therefore is a fundamental part of all matter. The electron is the smallest particle of electricity. It is the naturally occurring unit of electricity that exists in all matter. The simple atomic theory may also be regarded as the electron theory of matter. This electron theory may be used to explain all the electrical phenomena that either result from electrons in motion or because an excess or deficiency of electrons occurs in a body at a certain instant in time.

The coulomb
We have seen that the naturally occurring unit of electricity in all matter is the electron and for most practical purposes the electric charge of an electron is too small an amount in which to measure a quantity of electricity.

The unit of quantity of electricity or electric charge is the *coulomb*. It is the SI unit of electric charge and is defined as the quantity of electricity transported in one second by a current of one ampere.

The symbol used for quantity of electricity or electric charge is Q. The unit symbol used for coulombs after numerical values of charge is C.

**Current-flow
Convention**

Before the mechanism of current flow in solid conductors had been explained by the electron theory it was decided that the current flowed from the positive terminal of the battery through the external circuit to the negative terminal. This has remained the conventional direction of current in the external or load circuit. It is an unfortunate choice since it is the opposite direction to that of the electron drift through the external circuit.

An electric current is a flow of electric charges along a path in space defined by the run of the conductors. The current may consist of either
1 a flow of positive charges in one direction
2 a flow of negative charges in the opposite direction, or
3 a flow of positive and negative charges in opposite directions

The first possibility was chosen by the *conventional* current flow, although in fact the positive charges or ions are fixed in position in the crystal lattice, and therefore cannot move.

The second and third possibilities do occur since the electrons are free to move in solid conductors, and the positive and negative ions are mobile in liquid conductors or electrolytes.

From this it can be seen that the choice of the first possibility was unfortunate since the conventional current of positive charges is imaginary. However, this choice was made before the electron was discovered, and must be adopted, since all reference to electrical current direction in this and other textbooks means the *conventional* current direction.

If the current direction through the circuit is the same at any instant in time it is known as a *direct* current.

If the current direction changes periodically it is known as an *alternating* current.

The ampere

An electric current is the movement of a quantity of electricity along a path in space. It is measured by the rate of flow of electric charge. If one coulomb passes a given point in the path in one second the rate of flow is one coulomb per second. This is unit current flow and it is termed one ampere.

If Q coulombs pass a given point in the path of t seconds then the current I amperes is given by the expression:

$$I = Q/t \text{ amperes} \qquad (4.1)$$

where $I =$ steady current in amperes

$Q =$ quantity of electricity, or electric charge in coulombs

$t =$ time taken in seconds for the transfer of the charge to pass the point in the circuit

In the International System of Units an ampere is defined as the current that flows in each of two infinitely long parallel conductors situated one metre apart in a vacuum and producing a force of 2×10^{-7} newton per metre length on each conductor.

The conductors are attracted to each other when the currents flow in the same direction, but repel each other when the currents flow in opposite directions.

This definition of the ampere was adopted by the International Electrotechnical Commission in 1950 to form the basis of the definitions of other electrical and magnetic units.

The ampere as defined above is also one of the six basic units adopted by the General Conference of Weights and Measures at its tenth meeting in 1954 in Paris, to serve as the basis for the establishment of a practical system of measurement for international purposes. At the eleventh meeting, in 1960, the General Conference of Weights and Measures named this system as the International System of Units and decided on its international abbreviation of SI. The following table gives the basic six SI units, their names and their unit symbols.

Quantity	Unit name	Unit symbol
Length	metre	m
Mass	kilogramme	kg
Time	second	s
Electric current	ampere	A
Absolute temperature	degree kelvin	K
Luminous intensity	candela	cd

Current is measured by an instrument called an ammeter.

The symbol for steady current is I

The symbol for instantaneous current is i

The unit of current is the ampere. The abbreviation of the unit after numerical values is A. For example,

30 amperes may be written 30 A

20 milliamperes (20×10^{-3} A) as 20 mA.

Note the use of the prefix m for the abbreviation of the sub-multiple 10^{-3}. For a list of abbreviations of multiples and sub-multiples, see page 11.

Electric circuits

If an electric current is to flow, there must be a continuous conducting path for it to pass along so that the current can transfer electrical energy from one point to another in the path. The complete continuous path is known as an electric circuit, and the current acts as the carrier of electrical energy around this circuit.

All electric circuits have three essential parts:

1 the source

2 the conductors between source and load

3 the load

The source is that part of the circuit which produces the electrical energy at the expense of other forms of energy (chemical, mechanical and thermal).

The conductor system is the part of the circuit between the source and the load. It consists of insulated metal conductors in the form of cables and switches or switchgear.

The load is the part of the circuit that utilises the electrical energy by reconverting it to chemical, mechanical or thermal energy.

The path may be made discontinuous by opening a switch in the circuit. This action inserts an insulator—usually air or a mineral oil—in the conducting path. The insulator stops the current flow and the transfer of electrical energy from the source of the load.

The speed of electrical energy transfer is approximately that of light (186 000 miles per second or 3×10^8 metres per second). The speed of electron drift round the circuit is small by comparison with the speed of energy transfer.

Circuit diagrams

Electrical circuits may be simply represented in circuit diagrams. The conductors or wires of the circuit are shown by straight lines drawn parallel to the edges of the paper and standard graphical symbols are adopted for the various electrical apparatus. Some of these symbols are given in Fig. 4.1.

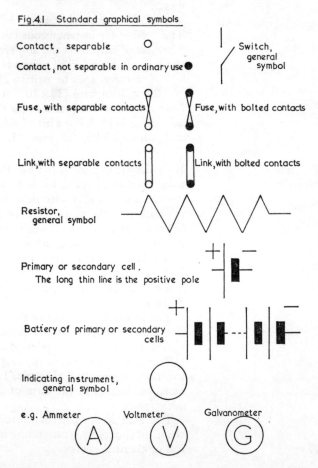

Fig.4.1 Standard graphical symbols

Contact, separable

Contact, not separable in ordinary use

Switch, general symbol

Fuse, with separable contacts

Fuse, with bolted contacts

Link, with separable contacts

Link, with bolted contacts

Resistor, general symbol

Primary or secondary cell.
The long thin line is the positive pole

Battery of primary or secondary cells

Indicating instrument, general symbol

e.g. Ammeter   Voltmeter   Galvanometer

The simple electric circuit of Fig. 4.2 uses some of these standard symbols to depict a d.c. source of a simple cell supplying a d.c. load comprising a resistor via a switch S1 and two connecting wires *ac* and *bd*.

Fig.4.2

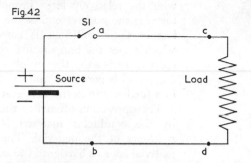

Use of an ammeter

If the current flowing in a circuit is to be measured an instrument called an ammeter is used. An ammeter indicates the amount of current either in amperes or multiples or sub-multiples of amperes on a scale provided on the instrument. The scale depending on the range of the instrument is graduated either in amperes or multiples or sub-multiples of amperes.

To measure an electric current the ammeter is so arranged in a circuit that the current passes through the instrument. This is done by 'breaking' the circuit and inserting the ammeter in the break to recomplete the circuit. The ammeter is then 'in series' with the load through which the measured current flows.

Fig. 4.3 shows how an ammeter is connected into the circuit of Fig. 4.2 to determine the magnitude of the current I that flows in the circuit when the switch S1 is closed.

Fig.4.3

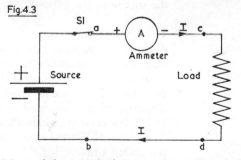

The effects of electric current

Most of the practical uses and benefits of electricity are covered by the following three effects produced by an electric current. They are

1 heating effects
2 magnetic effects
3 chemical effects

The heating effect

It was discovered that a conductor became heated whenever an electric current flowed through it.

This effect is caused by the collision of the electrons with the relatively large immobile ions as they drift towards the positive end of the conductor. The energy lost in these collisions is converted to heat energy, which raises the temperature of the conductor. If the current increases, the number of collisions will increase. This generates more heat energy, which results in a further rise in the temperature of the conductor.

The opposition offered to the axial drift of electrons by the conductor material is termed the *electrical resistance* of the material. The immobile ions in the material act as a barrier to the axial drift of the electrons in the conductor.

The heating effect of electrical current is usefully employed in fires, filament lamps, cookers, kettles, irons and so on, and in the safety fuse where abnormal current provides sufficient heat to melt the fuse wire. However, the heating effect can be a problem since the heat losses in conductors mean less efficient transmission and generation of electrical energy. If the conductors in cables or electrical appliances carry a greater current than their design permits, the heating losses will cause dangerous over-heating and this may result in fires in the cables or appliances. The fire risk is a major disadvantage of the heating effect of an electric current.

The direction of the electric current does not matter to the heating effect, since heat energy is produced whether the current flows in the forward or reverse direction. The heating effect of the current depends on the square of the current.

The magnetic effect

A compass needle or small magnet will deflect if it is placed near a conductor through which electric current is flowing. If the direction of the current is reversed, the direction of the needle's deflection will change. The magnetic effect is detected as long as the current flows, and if the current is increased the deflection increases. This shows that in the region around the wire the current has set up a magnetic field, whose direction depends on the current's direction and whose strength depends on the magnitude of the current. The shape of the magnetic field depends on the geometrical arrangement of the current-carrying conductors.

Practical applications of this effect are found in electric bells, relays, electromagnets, and so on.

The chemical effect

When a current passes through solutions of some chemical compounds, chemical changes take place. These changes depend on the chemical composition of the conducting solution or electrolyte, and the manner in which the current enters and leaves it—that is, the nature of the positive and negative electrodes.

This effect is usefully employed in electroplating, the operation of primary and secondary cells, and the production and refinement of some chemicals.

Corrosion of some metals is accelerated by this action.

Faraday discovered that when a current was passed between two copper plates immersed in a copper-sulphate solution, copper was deposited on the negative plate—that is, the cathode—and that the amount of copper deposited was proportional to the quantity of electricity passed through the electrolyte. This process is called electrolysis. It can be defined as the production of chemicals of dissociation of the electrolyte and inter-action at the electrodes, caused by the passage of an electric current through the electrolyte.

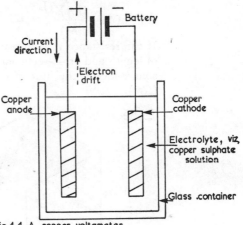

Fig.4.4 A copper voltameter

A voltameter (see Fig. 4.4) is an electrolytic cell arranged to measure the products of electrolysis.

Note: This apparatus must *not* be confused with a voltmeter, which is an instrument that measures voltage.

Dependence of resistance on dimensions and materials

If a conductor of uniform cross-sectional area is made of a homogeneous material—that is, of the same material throughout its length and cross-section—experiments show that the resistance (R) of the conductor is directly proportional to its length (l) and inversely proportional to its cross-sectional area (a) provided the temperature of the conductor remains constant. In symbols

$R \propto l$ if temperature constant

$R \propto l/a$

$\therefore R \propto l/a$

$$R = \frac{\rho l}{a} \text{ (ohms)} \tag{4.2}$$

The constant of proportionality ρ (the Greek letter rho) depends on the material of the conductor. It is an electrical constant for the conductor material at a fixed temperature and is known as the resistivity or the specific resistance of the material (see page 42).

The resistivity of a material is the resistance of a piece of the material of unit length and unit cross-sectional area. This definition is obtained by substituting unit length and unit area in equation 4.2. An alternative definition is: Resistivity of a material is the resistance between opposite faces of a unit cube of the material (see Fig. 4.5).

From equation 4.2

$$\rho = \frac{Ra}{l}$$

units of $\rho = \dfrac{\Omega \times \cancel{L}^{2}\,\text{L}}{\cancel{L}\,1} = \Omega\text{L}$

If the resistance R is measured in ohms and the units of length and cross-sectional area are the metre and the square-metre, the unit of resistivity is the ohm metre (Ωm).

Fig.4.5

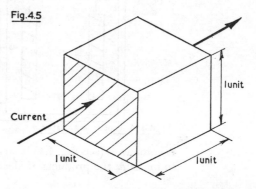

Current

1 unit

1 unit

1 unit

Effect of temperature on resistance

The resistance R_θ ohms of a material at any temperature degrees Celsius (formerly called Centigrade) may be predicted from the following expressions:

$$R_\theta = R_0(1 + \alpha\theta) \text{ ohms} \tag{4.3}$$

where R_0 = resistance in ohms of the material at zero degrees Celsius

    α = temperature coefficient of resistance per degree C

Let $R_0 = 1$ ohm and $\theta = 1°C$

and substitute in expression 4.3

then $R_1 = 1(1 + \alpha 1)$

$R_1 = 1 + \alpha$

$\alpha = (R - 1)$ per degree Celsius.

Thus the temperature resistance coefficient of a material may be defined as the increase in resistance in ohms of one ohm of the material at 0°C when the temperature of the material rises from 0°C to 1°C.

The temperature coefficient of resistance is α (alpha) per degree Celsius, which is abbreviated after a numerical value to per degree C.

The resistance of a material increases with a rise in temperature and decreases with a fall in temperature when its temperature resistance coefficient has a positive value. If a material possesses a negative temperature coefficient of resistance, its resistance decreases with a temperature rise and increases with a temperature fall.

When the value of a material's temperature coefficient of resistance is negligibly small, its resistance at any temperature is the same as that at zero degrees Celsius.

This means that the resistance of such a material is independent of variation in its temperature.

Pure metals have positive temperature resistance coefficients so that the resistances of conductors made from them increase as their temperatures rise. The exception is the material carbon, which has a negative temperature resistance coefficient so that its resistance decreases as temperature rises.

Resistors—which are conductors designed to have a definite degree of resistance—are usually wires made from special alloys. An alloy material is a mixture of metals which has different properties from its constituent metals. By varying the percentages of the constituent metals alloys are produced which have negligibly small temperature coefficients of resistance so that their resistance does not alter with temperature variations. Such alloys are used in the manufacture of standard resistors and heating elements; and examples of them are manganin, constaten.

Electrolytes, or liquid conductors, have negative temperature resistance coefficients, which means that their resistance falls with temperature rise.

Insulators are made from materials possessing very high electrical resistivity. They confine the flow of current to the circuit conductors by their insulating property. Unfortunately they also have negative temperature coefficients of resistance which means their resistance decreases as they grow hotter. This implies a deterioration of their insulating property with rise in temperature.

Voltage

Voltage is a term that is applied indiscriminately to the electro-motive force (e.m.f.) in a circuit and to the potential difference (p.d.) between two points in a circuit since both e.m.f. and p.d. are measured in volts.

Electromotive force (e.m.f.)
symbol: E;
unit: volt

Before an electric current will flow round a complete electrical circuit, somewhere in that circuit there must be an influence to move electric charges. This influence is known as the electromotive force, or in its abbreviated form as the e.m.f. The part of the circuit that provides the e.m.f. is called the source.

The e.m.f. is the energy given by the source to each unit quantity of electricity passing through the source. The SI unit of energy is the *joule* and the SI unit of quantity of electricity is the *coulomb*. Thus e.m.f. of the source = number of joules of energy/coulomb supplied by the source.

The SI unit of e.m.f. is the volt.

A *volt* may be defined as the e.m.f. of a source that gives one joule of energy to each coulomb of electric charge passing through it. A *voltmeter* is the instrument used to measure e.m.f. in volts.

Potential difference (p.d.)
symbol: V;
unit: volt

Potential difference between two points in a circuit is the energy in joules lost by each coulomb of electricity that passes between the two points. The convention of current flow assumes that positive charges travel from a point of higher potential to a point of lower potential. When a current flows between the two points in a circuit, the electrical energy taken from the current is converted into other forms of energy, namely into heat energy, light energy, mechanical energy or chemical energy.

As a coulomb of electricity passes through a potential difference of V volts, it loses V joules of energy. It is a well-known fact that water will flow from a higher level to a lower level; similarly, an electric current will flow from a higher to a lower potential point in a circuit. The difference in level or pressure head in a water circuit may be compared with the difference in electrical pressure or potential in the electric circuit. For water to flow between levels there must be a path for it to flow along, and similarly with electricity there must be a continuous conducting path between two points for an electric current to flow. However, a potential difference can exist between two points in both the hydraulic and electric circuits without a flow of current between them. For levels, sea-level is usually taken as the reference level for datum, and for potentials of points, the earth potential is taken as datum. The positive and negative terminals of the d.c. source are the points of highest and lowest potential in an electric circuit. There is a drop in potential along the circuit in the direction from the positive terminal—that is, in the conventional direction of current.

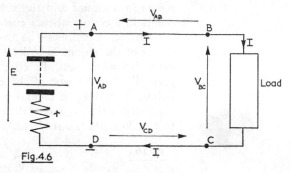

Fig.4.6

The symbol used for e.m.f. is E (see Fig. 4.6).

The symbol used for potential and potential difference is V.

The unit of both e.m.f. and p.d. is the volt, which is abbreviated after a numerical value to V. For example,

10 volts may be expressed as 10 V

10 000 volts may be written 10 kV

Note the use of the prefix to denote 1000

A useful notation for potential and potential difference is given below:

V_A = potential of A with respect to earth potential

V_B = potential of B with respect to earth potential

V_C = potential of C with respect to earth potential

V_D = potential of D with respect to earth potential

$$V_A \qquad - \qquad V_B \qquad = \qquad V_{AB}$$

Potential of A	$-$ Potential of B	= Potential differ-
with respect	with respect	ence between A
to earth	to earth	and B or
		voltage drop
		from A to B

Thus V single suffix refers to the potential of a specific point with respect to earth potential,

and V double suffix refers to the potential difference between the two points.

Note also in the diagram of Fig. 4.6 that the arrow heads indicating current direction show that the current I flows from a point of higher positive potential to one of a lower positive potential. Another convention used in Fig. 4.6 is that the arrow heads on arrows indicating e.m.f., voltage drops or p.d.s between any two terminals always point towards the terminal with the higher positive potential.

A voltmeter is the instrument used to measure p.d. or voltage drop in volts.

Use of a voltmeter

A voltmeter is an instrument for measuring voltage and is provided with a scale graduated either in volts or multiples or sub-multiples of volts. It is used to measure e.m.f. of a source; the potential difference between two points in a circuit and the voltage drop between two points in a circuit when a current flows between those two points. A voltmeter is connected across the two points of the circuit in order to measure the p.d. that exists between them. In effect the instrument is 'in parallel' with the part of the circuit between the two points. Therefore, it should possess a high resistance compared with that part of the circuit it bridges in order that it does not 'short circuit' the two points.

Fig.4.7a

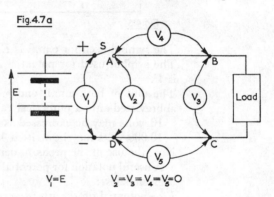

$$V_1 = E \qquad V_2 = V_3 = V_4 = V_5 = 0$$

Fig.4.7b

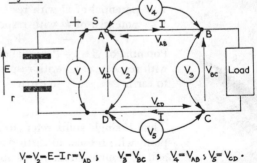

$$V_1 = V_2 = E - Ir = V_{AD} \; ; \qquad V_3 = V_{BC} \; ; \qquad V_4 = V_{AB} \; ; \qquad V_5 = V_{CD} \, .$$

Consider the circuit shown in Fig. 4.7a with switch S open and with the voltmeter connected in turn to the five positions indicated as V_1, V_2, V_3, V_4, and V_5. The voltmeter in these positions reads as follows:

V_1 gives the p.d. between the battery terminals when the battery is on open circuit or no load.

V_1 also equals the e.m.f. of the battery E volts in this instance since no current flows through the internal resistance of battery to produce an internal voltage drop.

V_2, V_3, V_4 and V_5 are all zero. This is because all the points A, B, C and D are at the same potential so that no potential difference exists between them to cause current flow.

When switch S is closed as shown in Fig. 4.7b the voltmeter readings V_1 and V_2 will be the same but less than the previous V_1 reading of the circuit of Fig. 4.7a, that is, less than the e.m.f. E by the internal voltage drop of the battery.

Reading V_3 will nearly be the same as reading V_2, their difference $(V_2 - V_3)$ if any is the voltage drop across the wires AB and CD. Reading V_4 gives the voltage drop across the wire AB and reading V_5 the voltage drop across the wire CD when the load current flows through the wires.

Thus $V_2 - V_3 = V_4 + V_5$

If the voltage drop in the wires is negligibly small, that is,

if $\quad V_4 + V_5 = 0$

then $\quad V_2 - V_3 = 0$

$\therefore \qquad V_2 = V_3$

This means that if the voltage drop in the conductors can be neglected then the terminal voltage of the battery on load equals the voltage drop or p.d. across the load.

Note: With d.c. voltmeters used in d.c. circuits the the polarity of the instrument must be correct for it to read up-scale.

Ohm's law

By experiment it has been found that provided the physical conditions of a conductor remain constant, the potential difference across the ends of most conductors is proportional to the steady current passing through them. This experimental relationship is known as Ohm's law.

Ohm's law states that as long as the physical conditions—that is, temperature, length and cross-sectional area—of the conductor are unchanged, the ratio of the potential difference (V) between the ends of the conductor to the steady current (I) flowing in the conductor is constant.

In symbols

$$\frac{V}{I} = \text{constant} = R$$

The constant R is the electrical resistance of the conductor.

If $\quad V = $ p.d. across the conductor in volts

and $\quad I = $ steady current through the conductor in amperes

then $\quad R = $ resistance of the conductor in ohms.

The unit of resistance is the ohm.

An instrument that measures resistance in ohms is termed an *ohmmeter*.

The symbol for the abbreviation of the unit ohm after numerical values is the Greek capital letter omega (Ω).

Resistors that obey Ohm's law are known as linear resistors since their volt/ampere characteristics are straight lines (see Fig. 4.8). An example is a metallic conductor with a definite resistance at a constant temperature.

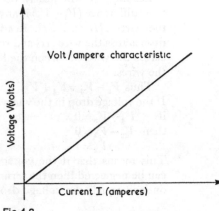

Fig.4.8

Resistors are known as non-linear resistors if their volt/ampere characteristics are not straight lines. Semiconductors are examples of non-linear resistors.

Ohm's law for a part of a circuit may be expressed in any of the following ways:

$$V = IR \text{ volts} \tag{4.4}$$

$$I = \frac{V}{R} \text{ amperes} \tag{4.5}$$

$$R = \frac{V}{I} \text{ ohms} \tag{4.6}$$

where V = p.d. in volts across the ends of a conductor or across a part of a circuit

I = current in amperes flowing through the conductor or part of the circuit

R = resistance in ohms of the conductor or the part of the circuit.

If the algebraic process of changing the subject of the equations just given is difficult for the student, the following method for memorising the three equations may be employed:

1 Sketch Fig. 4.9.

2 Cover the symbol of the quantity required

The symbols of the other quantities will then appear in the correct order to give the quantity required.

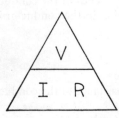

Fig.4.9

If V is required and covered in the diagram, IR appears uncovered.

If R is required and covered in the diagram, $\dfrac{V}{I}$ appears uncovered.

If I is required and covered in the diagram, $\dfrac{V}{R}$ appears uncovered.

Ohm's law for a complete circuit may be expressed in symbols in any of the following ways:

$$E = IR \text{ volts} \tag{4.7}$$

$$I = \frac{E}{R} \text{ amperes} \tag{4.8}$$

$$R = \frac{E}{I} \text{ ohms} \tag{4.9}$$

where E = resultant e.m.f. in volts in the complete circuit

$\quad\quad I$ = current in amperes flowing around the complete circuit

$\quad\quad R$ = total resistance in ohms of the circuit.

Again the previous method for memorising these equations may be used, but with the following modified diagram in Fig. 4.10.

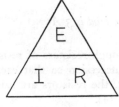

Fig.4.10

Examples on Ohm's law

Fig.4.11

Example 1

If a direct current of 12 A flows through a 9-Ω resistor what potential difference exists between the ends of the resistor?

By Ohm's law

$\quad\quad V = IR \text{ volts}$

where $\quad I = 12$ A

$\quad\quad\quad R = 9 \text{ }\Omega$

$\therefore \quad\quad V = 12 \times 9 = 108$

$\quad\quad\quad V = 108 \text{ volts}$

$\therefore \quad$ *P.D. between the ends of the 9-Ω resistor is 108 V*

Example 2

A circuit takes a current of 20 mA when connected across a 240 V d.c. supply. Calculate the resistance of the circuit.

By Ohm's law

$\quad\quad R = V/I \text{ ohms}$

where $V = 240$ V

Fig.4.12

$$I = 20 \text{ mA} = \frac{20 \text{ A}}{1000} = 20 \times 10^{-3} \text{ A} = 0 \cdot 02 \text{ A}$$

$$\therefore \quad R = \frac{240}{0 \cdot 02} = 12\,000$$

$$\text{or} \quad R = \frac{240}{20/1000} = \frac{240}{20} \times 1000 = 12\,000$$

or $\quad R=\dfrac{240}{20\times10^{-3}}=\dfrac{240}{20}\times10^3=12\,000$

$\therefore \qquad R=12\,000\ \Omega$

$\qquad\qquad R=12\ k\Omega$

$\therefore \qquad$ *Resistance of the circuit is 12 kΩ*

Example 3

A d.c. load with a resistance of 8 Ω has a p.d. at its terminals of 24 V. Determine the current flowing through the load.

By Ohm's law

$\qquad\qquad I=V/R$ amperes

where $V=24$ V

$\qquad\qquad R=8\ \Omega \quad \therefore I=24/8=3$

$\qquad\qquad\qquad I=3$ A

$\therefore \qquad$ *Current flowing through the load is 3 A*

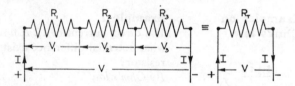

Fig.4.13

Resistors in series

If a current I amperes flows through resistors of resistance R_1, R_2 and R_3 ohms in succession so that each resistor carries the same current, the resistors are said to be connected in series (Fig. 4.14).

Fig.4.14

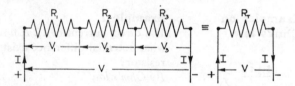

If a single resistor of R_T ohms produces the same current I amperes for the same total potential difference V volts applied, then

$\qquad R_T=R_1+R_2+R_3 \qquad\qquad\qquad$ **(4.10)**

Thus the total equivalent resistance of the three unequal resistors connected in series is the sum of the individual resistances. For any number (n) of unequal resistors connected in series:

$\qquad R_T=R_1+R_2+R_3+\ldots\ldots+R_n$

The total resistance of n unequal resistors connected in series is the sum of the individual resistances of the resistors.

Series law for *n* unequal resistors in series

$R_T=R_1+R_2+R_3+\ldots\ldots+R_n \qquad\qquad$ **(4.11)**

Series law for *n* equal resistors in series

If $R_1=R_2=R_3=R_n=R$

$\qquad$ then $R_T=nR \qquad\qquad\qquad\qquad\qquad$ **(4.12)**

The total resistance of n equal resistors in series is equal to the product of the number of resistors in series and the resistance of one resistor.

If the resistors R_1, R_2 and R_3 ohms are connected between the points A and B as shown in Fig. 4.15 they provide separate paths into which the main current (I) divides, and each resistor has the same potential difference across it; that is, V_{AB}. The resistors are said to be connected in parallel.

Fig 4.15

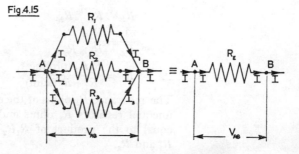

If R_E ohms is the resistance of a resistor equivalent to the three resistors in parallel, then

$$\frac{1}{R_E} = \frac{1}{R_1} + \frac{1}{R_2} + \frac{1}{R_3} \tag{4.13}$$

Thus for three resistors in parallel the reciprocal of the resistance of the equivalent resistor is equal to the sum of the reciprocals of the resistances of the three branch resistors.

Parallel law for n unequal resistors in parallel
In general, for any number (n) of unequal resistors connected in parallel the reciprocal of the resistance of the equivalent resistor is equal to the sum of the reciprocals of the branch resistances:

The reciprocal of resistance; that is $\dfrac{1}{\text{resistance}}$ is called conductance, and its symbol is G. The SI unit of conductance is the siemens. The abbreviation of the unit of conductance after numerical values is S.

Thus for resistors in parallel, the conductance of the equivalent resistor is equal to the sum of the conductances of the branch resistors.

$$G_E = G_1 + G_2 + G_3 + \ldots \ldots + G_n \tag{4.14}$$

Parallel law for n equal resistors in parallel
If $R_1 = R_2 = R_3 = R_n = R$ ohms

then $\dfrac{1}{R_E} = \dfrac{n}{R}$

$$\therefore \quad R_E = \frac{R}{n} \text{ ohms} \qquad (4.15)$$

For n equal resistors in parallel the resistance of the equivalent resistor is equal to the resistance of any branch resistor divided by the number of branches in parallel.

Parallel law for two unequal resistors in parallel

If resistors of R_1 and R_2 ohms resistance are connected in parallel then

$$\frac{1}{R_E} = \frac{1}{R_1} + \frac{1}{R_2}$$

$$\frac{1}{R_E} = \frac{R_2 + R_1}{R_1 \times R_2}$$

$$\therefore \quad R_E = \frac{R_1 R_2 \text{ (product)}}{R_1 + R_2 \text{ (sum)}} \qquad (4.16)$$

The resistance R_E ohms of the equivalent resistor of two unequal resistors R_1 ohms and R_2 ohms in parallel is equal to the product of $R_1 R_2$ divided by the sum of R_1 and R_2.

Electrical power and energy

When an electric current of I amperes flows through a conductor of resistance R ohms, the electric power P watts dissipated in the conductor is given by the expression.

$$P = I^2 R \text{ watts} \qquad (4.17)$$

Also by Ohm's law the voltage drop V volts across the conductor is

$$V = IR \text{ volts}$$

Since

Power $P = (IR) \times I$ watts

therefore

$$P = VI \text{ watts} \qquad (4.18)$$

Also

Power $P = (IR) \times \dfrac{(IR)}{R}$ watts

therefore

$$P = \frac{V^2}{R} \text{ watts} \qquad (4.19)$$

From the expressions 4.17, 4.18, and 4.19 power in watts dissipated in a circuit can be calculated if the values are known for current I amperes, resistance R ohms and voltage V volts applied to a circuit.

If the power to kilowatts is required the above values of the power in watts are divided by a 1000 since

$$1 \text{ kW} = 1000 \text{ W}$$

If the power P watts is supplied for t seconds, then P $\times t$ joules or $I^2 Rt$ joules is the electrical energy converted to heat energy by the the electrical resistance R ohms of the circuit appliance or conductor.

If the power is measured in kilowatts and the time in hours, then the electrical energy is measured in kilowatt hours.

$$1 \text{ kWh} = 1000 \text{ watts} \times 3600 \text{ seconds}$$
$$= 3\ 600\ 000 \text{ watt-seconds or joules}$$
$$1 \text{ kWh} = \text{Board of Trade unit of electrical energy.}$$

One kilowatt-hour is the Board of Trade unit of electrical energy sometimes referred to as the unit of electricity. The cost per unit of electrical energy is the charge made for each kilowatt-hour by the Area Electricity Boards to the consumer.

Exercises

1 Complete the following sentences:
    a the SI unit of electric charge is the . . . .
    b an . . . . . . is the instrument used to measure current
and **c** the SI unit of current is the . . . .

2 What is an electric current? Define the unit of electrical current.

3 Give the current flow convention and illustrate your answer with a suitable diagram.

4 Name the three effects of an electric current. How do these effects depend on the direction of the current?

5 How does the resistance of a conductor depend on its dimensions and the conductor material?

6 Define the temperature coefficient of resistance of a material. What effect does variation in temperature have on the resistance of conductors and insulators?

7 Give the three different effects of an electric current. For each effect give an example of a piece of apparatus in which the particular effect is used. Describe briefly the operation of one of your examples.

8 Explain the terms 'potential difference' (p.d.), 'voltage drop', and 'electromotive force' (e.m.f.). In what units is each quantity measured?

9 How is a voltmeter used to measure voltage? Illustrate your answer with a diagram of a simple two-wire circuit.

10 State Ohm's law in words and symbols. Determine the current taken by a 40 Ω resistor from a 240 V d.c. supply.

11 The current taken by a heating element when connected to a 240 V d.c. supply is 6 A. Calculate the resistance of the element.

12 Three resistors of resistance 3 Ω, 4 Ω and 12 Ω respectively are connected **a** in series and **b** in parallel. What is the total resistance of each combination of resistors?

13 What current flows through the heating element of a 240 V electric iron rated at 720 W?

14 A cable has a total resistance of 0·32 Ω and carries a load current of 200 A. For this cable calculate:

      a its voltage drop

      b its power loss

and **c** the energy loss over a 24-hour period

5
Magnetism

Magnetic fields

A magnetic field is the space around a magnet or current-carrying conductors in which the forces due to the magnet or the current can be detected. Since a magnet is mentioned in this definition of a magnetic field, it is important to understand what is meant by a magnet.

Magnets

A magnet is a substance that possesses the following properties:

1 If suspended horizontally a magnet will point approximately in the direction of geographical or true north and south after the oscillations have stopped. The end of the magnet pointing towards the earth's magnetic north pole is the north-seeking end of the magnet. It is known as the north pole of the magnet. The other end of the magnet points towards the earth's magnetic south pole. It is the south-seeking end of the magnet which is called the south pole of the magnet.

This is a very important property of a magnet, since upon it depends the navigational device known as a compass needle.

2 Like magnetic poles (two north poles or two south poles) repel one another and unlike magnetic poles (a north and a south pole) attract each other.

3 The poles of a magnet are of equal strength. One pole cannot be obtained without producing an equally strong pole of the opposite kind.

Fig.5.1

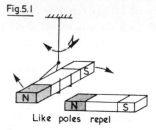

Like poles repel

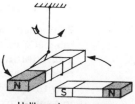

Unlike poles attract

4 Either pole of a magnet will attract substances such as unmagnetised iron, nickel, cobalt, their alloys and some of their compounds, especially lodestone, which is the naturally occurring magnetic oxide of iron. Substances that behave in this manner are known as magnetic substances.

5 A magnet is not weakened after it has been used to lift iron but it can lose its magnetism if subjected to violent treatment or is heated to red heat (800°C). The magnet will retain its strength indefinitely if carefully looked after and made of a suitable steel. Such magnets are known as permanent magnets. If not treated properly, however, it will weaken with time.

6 Pieces of soft iron that are attracted by a pole of a magnet temporarily acquire the magnet's property of attraction. The soft iron pieces seem to have poles. The ends of the iron near the pole of the magnet have poles of opposite kinds to that of the magnet. This phenomenon is known as magnetic induction.

Fig. 5.2a shows two soft iron nails attached to the north pole of a magnet. The nails repel each other at their lower ends, indicating that like poles have been induced in them.

Fig.5.2a

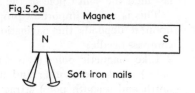

Magnet

Soft iron nails

Fig. 5.2b shows the lower ends have north poles since the north pole of another magnet repels them. When the soft iron nails are taken from the magnet, they no longer repel or attract each other, which shows that they do not retain their magnetism.

Fig.5.2b

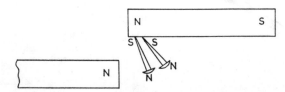

7 Magnetic action is not affected by a barrier of non-magnetic material such as paper, wood, lead or air. If a sheet of iron is placed in the magnetic field, however, the magnetic action is reduced. Magnetic screening may thus be obtained by enclosing the object concerned in a soft iron cylinder.

Magnetic materials

Substances may be classified as magnetic or non-magnetic materials.

Magnetic materials are substances such as iron, which are strongly attracted by permanent magnets. They are sometimes referred to as ferro-magnetic materials and include, besides iron, nickel, cobalt, and steel. These materials can be magnetised, and made into magnets.

Non-magnetic materials are substances which are not attracted by magnets and include wood, brass, air, glass, lead and copper. These materials cannot be made into magnets.

Magnetic flux

The region around a magnet or current-carrying conductors acquires a peculiar state in which a number of effects may be detected. This state may be described if it is assumed that a magnetic field of flux actually exists in the space around the magnet or the current-carrying conductors. The magnetic field may be visualised as a pattern of lines of magnetic force representing the paths taken by the magnetic flux which leave a north pole and enter a south. In spite of its name the magnetic flux does not flow in a magnetic field. Neither the lines of magnetic force nor magnetic flux have a physical existence—their presence is only detected by the

Fig.5.3 Magnetic fields of permanent magnets

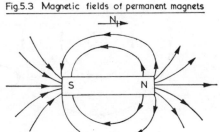

(a) N pole pointing North

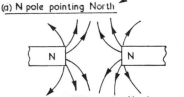

(b) N pole repelling another N pole

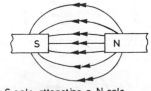

(c) S pole attracting a N pole

effects they produce. However, their assumed existence forms a useful basis for explaining the various magnetic effects and for calculating their magnitudes.

Properties of lines of magnetic flux

Although the existence of lines of magnetic flux is imaginary they appear to have the following properties:
1 Their direction at any point in a magnetic field set up in a non-magnetic medium is the direction indicated by the north pole of a compass needle placed at the point.
2 They form closed loops through the non-magnetic medium and the magnet or electromagnet.
3 They never intersect.
4 They are similar to stretched elastic cords in that they are always trying to contract.
5 They repel one another when they are parallel and in the same direction.

Some flux patterns of some simpler magnetic fields are given in Figs. 5.3, 5.4, 5.5, 5.6, and 5.7. These diagrams show the conventional relationships between the direction of flux, the magnetic polarity and the

Fig.5.4a Magnetic field due to a straight conductor carrying current.

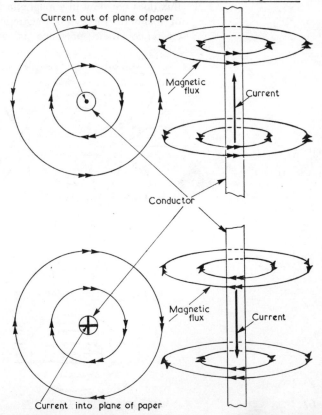

direction of current adopted in magnetism and electro-magnetism. It should be noted that there is no difference between the magnetic flux produced by permanent magnets and that produced by electric currents; that is, by electromagnetic means. The above properties apply to magnetic flux produced both by permanent magnets and by electromagnets.

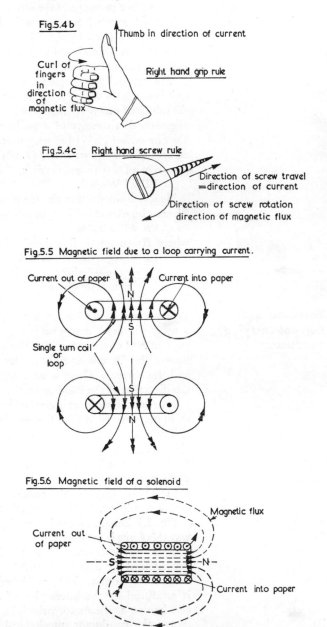

Fig.5.4 b

Thumb in direction of current

Curl of fingers in direction of magnetic flux

Right hand grip rule

Fig.5.4c   Right hand screw rule

Direction of screw travel =direction of current

Direction of screw rotation direction of magnetic flux

Fig.5.5 Magnetic field due to a loop carrying current.

Current out of paper   Current into paper

N
S

Single turn coil or loop

S
N

Fig.5.6 Magnetic field of a solenoid

Magnetic flux

Current out of paper

S — N

Current into paper

Magnetic flux density

The total amount of magnetic flux that passes through unit area at right angles to the direction of the magnetic flux is known as the magnetic flux density (or magnetic induction) for that part of the magnetic field. The greater the density of the magnetic flux, the stronger the field will be so that magnetic flux density may be regarded as a measure of the strength of a magnetic field.

The SI symbol for magnetic flux is Φ (the Greek capital letter phi). In the SI the unit of magnetic flux is the weber, which is abbreviated to Wb after numerical values of magnetic flux.

The symbol for magnetic flux density is B. Its SI unit is the tesla which is abbreviated to T after numerical values of magnetic flux density. One tesla (1 T) is the magnetic flux density of a uniform magnetic field that has one weber (1 Wb) of magnetic flux passing through an area of one square metre (1 m²) at right angles to the direction of the magnetic flux. Thus in symbols

$1T = 1$ Wb/m².

Hence, by definition the magnetic flux density in the SI may be expressed in symbols as

$B = \Phi/a$ teslas

where $B =$ magnetic flux density in teslas

$\Phi =$ magnetic flux in webers

$a =$ cross-sectional area in square metres of the magnetic circuit

Force on a current-carrying conductor

When a conductor carries a current at right angles to a magnetic field, a mechanical force is exerted on the current and consequently on the conductor. The direction of the force is at right angles to the magnetic field and the current. There is no component of the force in the direction of the conductor, that is, in the direction of the current. This force is created by the interaction of the main magnetic field with that produced by the current.

The magnitude of the force on a current flowing at right angles to a magnetic field is proportional to
1 the flux density of the magnetic field
2 the current
3 the active length of the conductor, that is, the length of the conductor in the magnetic field.

Fig. 5.7 illustrates the interaction of the field and the way this force is produced.

This phenomenon of a mechanical force being experienced by a current-carrying conductor is sometimes referred to as the law of interaction since the force is produced by the interaction of magnetic fields. It is also known as the motor principle, since when the force moves the conductor, mechanical energy is expended.

Fig.5.7

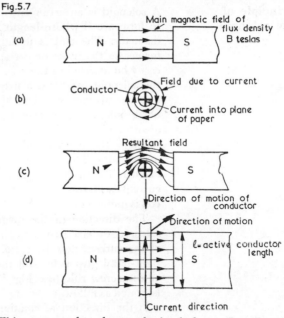

(a) Main magnetic field of flux density B teslas — N S

(b) Conductor — Field due to current — Current into plane of paper

(c) Resultant field — N S — Direction of motion of conductor

(d) Direction of motion — ℓ= active conductor length — N S — Current direction

This energy has been obtained from the electrical energy supplied to the arrangement by the electric current. Thus the assembly is a simple electric motor since it converts electrical energy to mechanical energy.

Fleming's left-hand rule for the motor principle
The directions of the main field, the current and the force of interaction are mutually at right angles. The direction of the force can be quickly found by Fleming's left-hand rule used as follows:

Hold the first two fingers and the thumb of the left hand at right angles to each other. Then, if the First finger points in the direction of the main Field or Flux, and the seCond finger points in the direction of the Current then the thuMb will point in the direction of the Motion of the conductor due to the force of interaction.

Note that the bold capitals serve as a memory aid. It is important that the correct hand is used for this rule. A useful memory aid to ensure that Fleming's Left-hand rule is used for the Motor principle is the fact that **L** and **M** are consecutive letters in the alphabet (see Fig. 5.8).

Fig.5.8

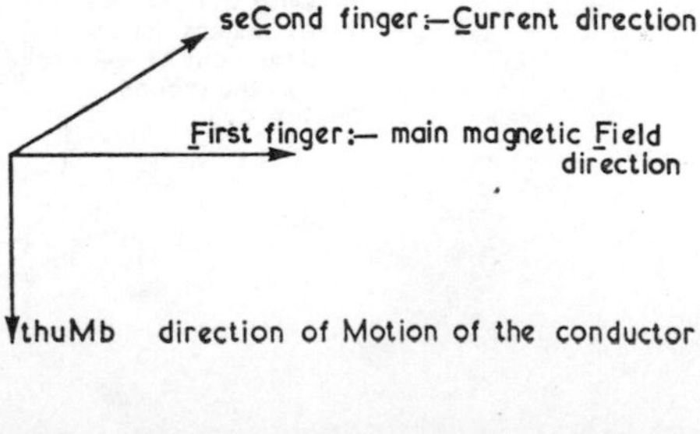

seCond finger:—Current direction

First finger:— main magnetic Field direction

thuMb direction of Motion of the conductor

**Principle of
the solenoid**

A solenoid is an arrangement of insulated wire wound to form a coil whose length is large compared with its diameter. The coil is usually wound on a cylindrical former so that rods or cores of different materials may easily be inserted in the coil. A current passing through the solenoid produces a magnetic field similar to that of a bar magnet; one end of the solenoid behaves as a north pole, and the other as a south pole. When a core of magnetic material such as iron is placed inside the solenoid, the magnetic effect of the solenoid is increased many times. This arrangement is known as an electro-magnet. The magnet effect also increases if the solenoid's current is increased, and if the number of turns in the coil is increased.

The direction of the magnetic field produced by a solenoid or electro-magnet is related to the direction of the current flow by certain conventions such as the right-hand grip rule, and the right-handed screw rule or corkscrew rule (see Fig. 5.9). This second rule may be stated as follows:

If the direction of rotation of a right-handed screw or corkscrew is the same as the direction of the current around the coil, then the direction of the magnetic field inside the solenoid is the same as the direction in which the screw or corkscrew travels.

Fig.5.9

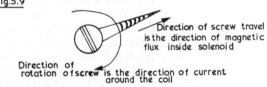

Direction of screw travel is the direction of magnetic flux inside solenoid

Direction of rotation of screw is the direction of current around the coil

The right-hand grip rule (see Fig. 5.10) states:

'If the right hand is closed so that the fingers curl in the direction of the current around the coil, then the outstretched thumb points in the direction of the magnetic flux inside the solenoid; that is, to the north end of the solenoid.'

Fig.5.10

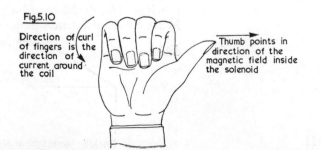

Direction of curl of fingers is the direction of current around the coil

Thumb points in direction of the magnetic field inside the solenoid

The magnetic polarity of the solenoid will change if either the current direction or the direction in which the coil is wound around the former is changed.

Uses of a solenoid

Solenoids or electromagnets are used in too many different ways for detailed descriptions of every application to be given here. However, it must be stated that without electromagnets to provide their magnetic fields, both a.c. and d.c. generators, motors, and many electrical measuring instruments would be unable to function.

Electromagnets also play an important role in the control and protection of electric circuits in the form of over-load earth-leakage and other protective relays. They are also vital in the field of telecommunications.

The only applications of the solenoid or electromagnet discussed in this section are in the manufacture of permanent magnets, in electric bells, buzzers, in the construction of a simple relay, in the telephone receiver and in indicator elements.

The manufacture of permanent magnets

The arrangement shown in Fig. 5.11a is suitable for making a permanent magnet. If a core of a hard steel is placed inside the solenoid and a current sent through the coil in the direction shown the steel will be magnetised, with the north and south poles as shown in Fig. 5.11a.

Fig.5.11a

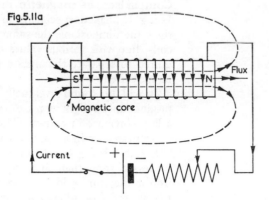

An alternative arrangement for making U-shaped magnets is shown in Fig. 5.11b. This arrangement also gives an approximate but useful method of comparing the strength of magnets and the magnetic properties of various materials.

The U-shaped specimen of material has two solenoids slipped over its limbs, and the current is arranged to flow through them so that they assist each other in magnetising the core. The spring balance measures

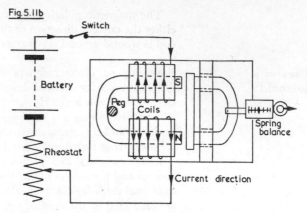

Fig.5.11b

the pull required to draw a soft iron armature away from the magnet. The reaction to the pull is taken by a non-magnetic peg holding the specimen in position on the board. This pull, read off the balance, is taken as a measure of the strength of the magnet.

By varying the current through the coils and the number of turns on the coils the pull required was found to be constant. This confirms that it is the product of the current and the number of turns—that is, the ampere turns—that is, the quantity producing the magnetism. This product is known as the magnetic-motive force.

Comparison of magnetic materials

Using U-shaped specimens of various materials with the same dimensions, the same number of turns on the coils, but with various values of current, the pulls required to separate the armature from the magnet may be measured.

It will be found eventually that each magnet will produce only a slight increase in magnetic strength against a large increase in current. The magnet is then said to be saturated or in a state of saturation.

Soft iron is found to be magnetised to the same extent by less current than that required by hard steel. Soft iron is therefore said to be more permeable than steel, that is, soft iron offers less opposition to being magnetised than steel.

Residual magnetism

If in the tests on each specimen the magnitude of the currents becomes such that the specimen is saturated, and the switch is then opened, the pull required to remove the armature from the magnet is a measure of the magnetism that remains in the magnet. This residual magnetism is known as the remanence of the material.

It is found that steel retains more magnetism than soft iron and that hard steel has a more remanent or residual magnetism than mild steel.

Another important property of a magnetic material is the relative ease with which the residual magnetism can be destroyed—that is, the magnet can be demagnetised.

This can be tested as follows:

After the specimen has been saturated and the switch opened, the connections to the coils are reversed. The switch is closed and a small current made to flow through the coils in the reverse direction while a small steady pull is exerted on the armature. The current is gradually increased until the armature is pulled away from the magnet by the small constant pull.

This value of current that demagnetises the material —that is, destroys its remanent magnetism—is a measure of the coercive force of the material. Hard steel has a higher coercive force than soft iron and thus hard steel opposes demagnetisation better than soft iron. Also, hard steel retains more residual magnetism than soft iron. These properties make hard steel (which has usually been alloyed with tungsten or cobalt) more retentive than soft iron, and thus more suitable for use as permanent magnets.

The more permeable and less retentive materials such as soft iron, mild steel, and iron silicon alloys, are used in electromagnets.

Effect of a non-magnetic material

If a thickness of paper is placed on each of the pole faces of the U-shaped specimen so that they do not touch the armature, the strength of the electromagnet is reduced considerably. This was shown by repeating the experiment and observing that the pull to part the armature from the magnet is very much less than it was without the layer of non-magnetic material. An air gap has the same effect, and this effect of any non-magnetic material is often referred to as the air gap effect. The strength of an electromagnet is thus greatly reduced by the introduction of an air gap into its magnetic circuit. If an air gap is unavoidable it is important that it be kept to a minimum.

Before leaving this section of the manufacture of magnets it is worth while to mention the simple explanation of the nature of magnetism in materials proposed in the *molecular theory of magnetism*.

Molecular theory of magnetism

Consider Fig. 5.12a which shows a group of five small compass needles. The small magnets have arranged themselves in a stable ring formation by the attraction of unlike poles. Fig. 5.12b shows the same five compass needles when the north pole of a strong bar magnet has

Fig.5.12a   Fig.5.12 b

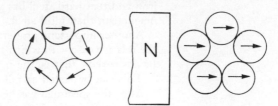

been brought near them. The north poles of the compasses are repelled by the strong north pole, and align themselves in the strong magnetic field of the bar magnet.

In the molecular theory a material in its unmagnetised state is assumed to possess countless stable ring formations of minute molecular magnets. This is a reasonable assumption since if a bar magnet is broken up into small pieces, the pieces themselves are magnets. The magnetic flux does not prefer any particular portion. If the process of subdividing the magnet is repeated until the magnet is finally reduced to its component molecules, the molecules themselves are assumed to behave as minute magnets. In the unmagnetised state these molecular magnets are arranged in stable ring formations very similar to that of the small compass needle magnets in Fig. 5.12a. When the material is saturated magnetically, then all the molecular magnets have arranged themselves in line with the magnetising field in the same way as the small compass needles in Fig. 5.12b. The molecular theory of magnetism is incomplete because it fails to explain how the molecular magnets receive their magnetism. The *current ring theory* answers this question, but will not be discussed at this stage.

However, the molecular theory of magnetism gives a simple explanation for the conversion of a magnetic material from its unmagnetised state to its magnetised state, and the general shape of a magnetisation curve for a magnetic material.

Consider Fig. 5.13a which shows the unmagnetised state of the magnetic material core of a solenoid. The molecule magnets are arranged in the stable ring formation before the solenoid has been energised—that is, before the current through the coil sets up the magnetising field. Fig. 5.13b shows the slight distorting effect of the magnetising field on the magnetic rings. Fig. 5.13c shows the material becoming more magnetised. Fig. 5.13d shows the material fully magnetised—that is, saturated. If the solenoid is de-energised at this stage the material returns to a magnetised state, somewhere between the conditions shown in Fig. 5.13b and 5.13c depending on the material. This is caused by the residual magnetism or remanence of the material.

The general shape of a magnetisation curve for a magnetic material is shown in Fig. 5.14. Portion ab corresponds to the transition from the unmagnetised state as shown in Fig. 5.13a to the slightly magnetised state as shown in Fig. 5.13b. Portion bc corresponds to the change in the magnetic state of the material from the condition shown in Fig. 5.13b to that shown in Fig. 5.13c. Portion cd represents the saturated state of the material. The magnetisation curve for a non-magnetic material is a straight line since here the ring formation of the molecular magnets is so stable that the magnetising field has no disruptive effect on them.

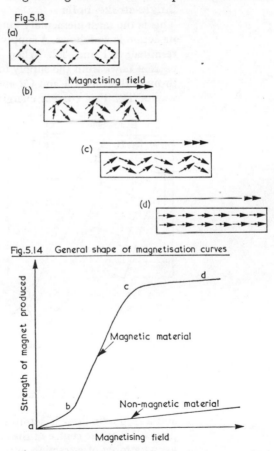

Fig.5.13

(a)

(b)   Magnetising field

(c)

(d)

Fig.5.14   General shape of magnetisation curves

Strength of magnet produced

Magnetic material

Non-magnetic material

Magnetising field

Electric bell

There are various types of electric bell including the single-stroke bell, trembler bell and continuously ringing bell, but all of them depend on the attraction exerted by an electromagnet on a soft iron armature.

Underdome bells

Most small modern bells are of the underdome variety. In this type of bell all the working parts are contained in the space under the gong. This form of construction makes a very compact unit.

Striker-arm bells

In larger bells and in the older types, the tendency in design is to keep the working parts separate from the gong or dome. Here the gong is reached by means of a striker arm, and this is the name we shall use for them. As the actions of this type of bell are easier to follow they will be used in the diagram to illustrate the various modifications of the basic types.

Single-stroke bells

This is the most elementary kind of bell (Fig. 5.15) and its action is as follows. A supply is connected across the terminals T1 and T2. When the circuit is closed a current flows from the supply to the bell by way of T1, through the coils C1 and C2 and back to the supply by way of T2. This current energises the coils C1 and C2

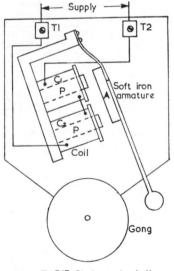

Fig 5.15 Single stroke bell

and a magnetic field is set up concentrated in the pole pieces P.P at the centre of the coils. This magnetic field exerts a force of attraction on the soft iron armature A. The armature moves against the action of the spring S, bringing with it the striker arm to hit the dome and emit a single note. The armature is held close to the ends of the pole pieces until the supply circuit is broken externally. Once the current ceases to flow the magnetic field collapses, releasing the armature. The spring then returns the striker arm to its original position. The fact that it only gives one note every time the supply is applied limits the usefulness of this type of bell. It is used for signalling purposes and a similar action is used for door chimes.

Trembler bells

In order to obtain a ringing sound that is maintained as long as the person trying to make contact keeps the supply circuit closed, a modified circuit is required. This will take a form similar to that shown in Fig. 5.16 and is called a trembler bell. It will be noted that the end of coil C1 is not connected directly to the supply terminal T1. Instead it is taken by way of the striker arm to a make-and-break contact arrangement, and then to the terminal T1.

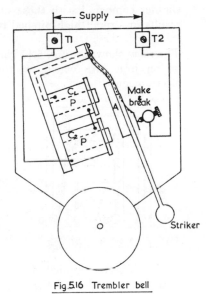

Fig.5.16 Trembler bell

The action of the bell is as follows. When a potential is applied to terminals T1 and T2 current flows from T1 through the make-and-break contacts to coil C1. From there it flows through coil C2 and back to the supply by way of T2. This causes a magnetic field to be set up, concentrated in the pole pieces, which attracts the soft iron armature and causes the striker to hit the gong. The movement of the striker arm breaks the circuit at the make-and-break contacts. This causes the magnetic field to collapse and allows the striker arm to return to its original position. As a result the circuit re-closes at the make-and-break and the entire process is repeated to give a second stroke. The bell will continue to ring until the supply to the bell terminals is interrupted.

The trembler bell is very widely used as a simple means of attracting attention in homes, offices, and hotels. It is also used as the audible component of warning systems such as those for critical water-levels, starting conveyor systems and so on. The only difference between the heavier industrial bell and the small domestic ones is that they may be designed for different supply voltages.

Buzzers

If a quieter signal is required than that given by a trembler bell, a buzzer is used. It is similar in construction to a trembler bell but it does not have a striker or gong. Its armature is also lighter and its movements slighter compared with those of a trembler bell. This results in a quiet high-pitched buzz.

A simple relay

The relay shown in Fig. 5.17 consists of two fixed coils on an iron core which together form a horse-shoe electromagnet.

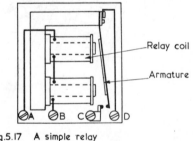

Fig.5.17   A simple relay

The coils are supplied via terminals A and B. Fitted close to the poles of the electromagnet is a spring-controlled armature. Very little power is required to attract the armature and to close the auxiliary contacts CD. There is no electrical connection between the circuits containing AB and CD, but a signal in a circuit containing A and B is relayed to a circuit containing C and D. Fig. 5.18 illustrates the way this type of relay is used in a simple bell circuit. When the push button—a switch that is normally open—is pressed, the battery is connected to the relay coil AB. The relay coil is thus energised. This attracts the armature of the relay, closing the normally open contacts CD. When the relay contacts are closed, the battery in the bell circuit is connected and the bell rings. If the push button is released, the relay coil de-energises. The armature springs back, opening the contacts CD in the bell circuit, and the bell stops ringing.

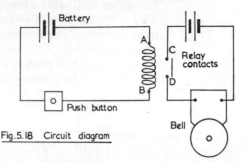

Fig.5.18   Circuit diagram

Telephone receiver

If a receiver is connected in series with a battery and a carbon microphone, it will convert the current variations produced by the carbon microphone—that is, the transmitter—back into sound. The receiver (Fig. 5.19) consists of a horse-shoe electromagnet formed by a permanent bar magnet fitted with two soft iron pole-pieces. These carry coils of thin insulated wire connected in series. The variations in current cause changes of

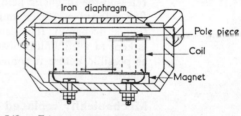

Fig.5.19   Telephone receiver

magnetic field strength which attract and release the flexible iron diaphragm. This sets up mechanical vibrations in the diaphragm producing sound variations identical with those transmitted by the carbon microphone.

A simple two-way speaking circuit needs two ordinary telephones and a battery all connected in series. Each telephone is fitted with a carbon microphone and a receiver.

Indicator elements

For a visual indication system to work a flag or disc must be moved from one position to another so that it will stand out against a group of such flags. When the change of position has been noted the flag must be returned to its original position. All standard forms of indicator elements cause the initial movement to take place by electromagnetic means. The differences between the various types lie in the methods used to restore the flag to its starting point. This can be done by electrical means, mechanical means, or by gravity.

Pendular indicator elements

This is the simplest of elements and for this reason the most common. The construction of a standard form is illustrated in Fig. 5.20. Its action is as follows. When the associated bell-push is pressed current flows in the coil. This sets up a magnetic field which draws the flag

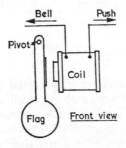

Fig. 5.20 Pendulum indicator

towards the coil and holds it there. When the push is released the magnetic field in the coil collapses and the flag swings freely back and forth on its supporting arm (hence its name). It is carefully balanced, and friction is kept to a minimum, so that it can swing for as long as possible before it comes to rest back in its starting place.

Mechanically replaced element

In this case the rear of the flag is pivoted at the reset rod and in the 'off' position the front end rests on the top of a movable soft iron armature (Fig. 5.21). When

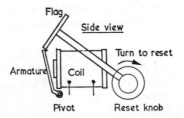

Fig. 5.21 Mechanically reset indicator

the coil is energised by its bell-push being pressed the armature moves up to the pole-piece of the coil. This displaces the flag, which falls into the second position shown in the diagram. It is replaced by rotating the reset arm against the action of a spring. Pegs on the reset arm catch on the body of the flag, and as long as the coil is no longer energised the flag can return to its place resting on the armature. Once the flag is back in position the reset arm can be released and the spring returns it to its former position.

Electrically reset element

Two coils are required for each element of this type. When the bell-push is pressed coil B (Fig. 5.22) is energised, moving the flag into position B. It remains in this position until it has been seen. A second reset push is then pressed which energises coil A and returns the flag to its starting point.

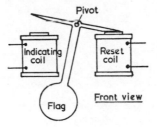

Fig. 5.22 Electrically reset indicator

Exercises

1 What is the difference between a magnetic material and a non-magnetic material? Sketch typical magnetisation curves for both types of materials.
2 State the properties of a magnet. What is **a** a temporary magnet, and **b** a permanent magnet?
3 Describe with the aid of a diagram an electrical method of making an unmagnetised steel rod into a magnet so that one end of the rod has a north polarity and the other end a south polarity. Show clearly on your diagram the current direction, the polarity of the steel rod and the direction of the magnetic flux.
4 Sketch the magnetic field set up by a solenoid. Indicate on your diagram the polarity at each end of the solenoid and the respective directions of the current in the coil, of the solenoid and the magnetic field. Give two practical applications of a solenoid.
5 Describe experiments to plot the magnetic fields due to **a** a bar magnet and **b** a straight current-carrying conductor.
6 A straight conductor carrying direct current is held lightly in a magnetic field at right angles to the direction of the field. Describe and explain with the aid of a

diagram why the conductor tends to move. Give two examples of the application of this effect.

7 Describe with the aid of a diagram or sketch the construction and operation of a trembler bell.

8 Draw a sketch or diagram of one form of relay used in electrical work. Explain its operation and with the aid of a diagram show the application of such a relay.

6
Sources of supply

It has been mentioned earlier in Chapter 4 that before an electric current will flow round a complete circuit somewhere in that circuit there must be an influence to move electric charges. This influence is known as the electromotive force (or e.m.f.). An e.m.f. may be obtained in several different ways, but it is generally produced by chemical means or by the method of electromagnetic induction. The part of the circuit that provides the e.m.f. is known as the source. The chemical sources of e.m.f. that will be discussed in this chapter are primary cells and secondary cells.

Production of e.m.f. by primary cells

In a primary cell electrical energy is produced from chemical energy. A chemical reaction releases chemical energy for conversion to electrical energy when the circuit external to the cell is closed; that is, when the load circuit demands a supply of electrical energy from the cell. The conversion of chemical energy to electrical energy in the case of primary cells is not reversible. Thus an exhausted primary cell cannot be recharged electrically. To recharge a primary cell the chemicals have to be renewed.

There is another class of cells known as secondary cells or accumulators that may be recharged electrically by passing a current through them in the reverse direction for a definite time.

The simple primary cell
A simple primary cell consists of two conducting materials, usually, in the form of plates or rods of dissimilar metals, known as electrodes, immersed in a conducting liquid or electrolyte. The electrodes form the positive and negative poles of the cell. The conducting materials of electrodes may be arranged in an electro-chemical series, and in the following list some of these materials are given in correct order, although certain other intermediate materials are omitted.
1 aluminium
2 zinc
3 iron
4 lead
5 hydrogen
6 copper
7 mercury
8 platinum
9 carbon

The further apart in the series the electrode materials are the greater is the e.m.f. produced by the cell. The list order from the negative to the positive pole is from top to bottom; thus for a simple cell with electrodes of copper and zinc, zinc is the negative pole and copper the positive pole.

The composition, concentration and temperature of the electrolyte also affect the magnitude of the e.m.f. produced by the cell. The cell's dimensions have no effect on the e.m.f. it produces, but they affect its internal resistance and capacity.

Dilute sulphuric acid is the electrolyte usually employed in a simple cell. When the external circuit is closed, the electrolyte reacts chemically with one of the electrodes, releasing hydrogen which collects as fine bubbles on the surface of the other electrode. This effect is known as *polarisation*. It reduces the e.m.f. and increases the internal resistance of the cell. Practical primary cells have a depolariser, which may be a solid or liquid chemical to remove the hydrogen. The action of a simple cell is shown in Fig. 6.1.

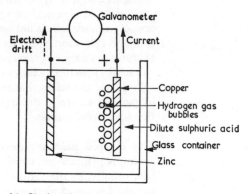

Fig.6.1 Simple primary cell

For current to flow around the complete circuit electric charges must pass through the electrolyte. The mechanism of current flow through a solid conductor has already been discussed in Chapter 3, page 40, and defined as the drift of electrons through the inter-spaces of the crystal lattice. In a conducting liquid or electrolyte the substance in solution divides into positive and negative electrically charged particles—in other words, ions. In dilute sulphuric acid hydrogen ions and sulphate ions are present, and the hydrogen ion is positive since it has a lost negative charge—an electron—while the sulphate ion is negative since it has gained electrons. When one molecule of sulphuric acid dissociates itself into ions, two positive hydrogen ions and one negative sulphate ion are produced.

This automatic division of a molecule into ions when in solution is known as dissociation. The movement of these positive and negative ions towards the electrodes constitutes a current through the electrolyte.

With pure zinc and copper electrodes, no chemical action takes place when the external circuit is open, although a d.c. voltmeter will show that an e.m.f. is present and that the zinc is negative with respect to the copper. The explanation is that positive zinc ions leave the zinc plate and enter the solution, creating a positively charged region around the zinc plate. This positive 'space charge' eventually prevents further positive zinc ions from leaving the plate by repelling them back into the plate. Thus the zinc plate is negatively charged since it has lost some positive zinc ions and is left with excess electrons.

When the external circuit is completed, these excess electrons drift along the external circuit from zinc to copper. The conventional current direction is thus from copper to zinc in the external circuit. This means that copper is the positive pole and zinc is the negative pole of the cell with respect to the external circuit.

Inside the cell the conventional current direction is from zinc to copper. Thus the positive ions, zinc and hydrogen, drift in this direction towards the copper electrode, and the negative sulphate ions drift towards the zinc electrode.

On arrival at the zinc electrode, the sulphate ion loses electrons which enter the external circuit, and the zinc and sulphate radical combine to form zinc sulphate, which enters the solution to dissociate immediately into zinc and sulphate ions. This chemical action, which involves gradual dissolution of the zinc, provides the energy to maintain the current flow.

At the copper electrode the positive hydrogen ions accept electrons to become hydrogen gas, which appears as fine bubbles on the copper electrode. This is the polarisation effect which reduces the e.m.f. and will increase the internal resistance.

As the hydrogen ions accept electrons from the anode the zinc ions entering the solution from the cathode give electrons to the external circuit.

Electrons accepted by the hydrogen ions at the anode from the external circuit are replaced by the electrons released by the zinc ions entering the electrolyte from the cathode, and the positive hydrogen ions lost from solution are replaced by the positive zinc ions. Thus the total number of electrons in the external circuit and the total number of positive ions in the electrolyte are constant. A primary cell appears to push electrons into the external circuit at the cathode, to replace the electrons pulled out of the external circuit at the anode.

Local action in a primary cell

If the cathode of a cell is made of commercial zinc it will dissolve in dilute sulphuric acid, liberating hydrogen when the cell's external circuit is open. The explanation is that commercial zinc contains impurities. When the zinc is placed in the electrolyte these impurities form the second electrodes of many small primary cells, with zinc as the other electrode. These small primary cells are short-circuited since both electrodes are in direct physical contact. The result is that currents flow and the chemical changes previously described take place. This action is known as local action of the cell.

If the zinc is rubbed with a small amount of mercury the mercury is absorbed by the zinc surface without reacting chemically with it. Thus the zinc amalgamates with the mercury to give amalgamated zinc. This prevents the local action from happening when the cell is not in use. It does not interfere with the normal action of the cell when the external circuit is closed, since the zinc ions pass easily through the mercury film on the surface of the zinc cathode.

This local action is responsible for the corrosion of metals especially in damp situations where dissimilar metals are in contact. (See IEE Regulations, 14th Edition, B41 to B44 and Appendix 3.)

There are many types of primary cell but the four types described in this section are
1 the Daniell cell
2 the Leclanché cell—wet type
3 the Leclanché cell—dry cell
4 the mercury cell

The Daniell cell

A Daniell cell is shown in Fig. 6.2. Its negative pole is an amalgamated zinc rod that stands in a porous pot and is amalgamated with mercury to prevent local action. The positive pole of the cell is the copper container. The porous pot contains dilute sulphuric acid and the copper container holds a saturated solution of copper sulphate. The solution of copper sulphate is kept at saturation point by the copper sulphate crystals held on a ledge inside the copper container.

In the cell both electrolytes ionise to provide positive hydrogen ions and negative sulphate ions in the dilute sulphuric acid solution in the porous pot, and positive copper ions and negative sulphate ions in the copper sulphate solution in the outer container. When the external circuit is completed the hydrogen ions pass through the porous pot into the copper sulphate solution, tending to travel towards the copper anode. The copper sulphate solution then has an excess of positive ions. However, the copper ions accept electrons from the copper anode more readily than the hydrogen ions do so that instead of hydrogen forming on the copper

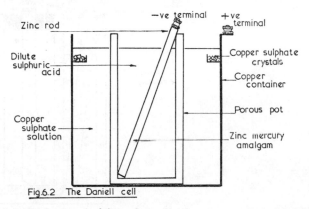

Fig.6.2   The Daniell cell

anode and polarising the cell, the copper ions having accepted electrons from the copper anode become metallic copper deposits on the copper anode. At the negative pole, zinc reacts with the dilute sulphuric acid to form zinc sulphate which appears in solution as positive zinc ions and negative sulphate ions. This releases electrons from the solution and passes them to the external circuit via the cathode.

The electrons drift round the external circuit to the anode where they neutralise the positive copper ions that appear there from the copper sulphate solution. This electron drift is the load current of the circuit.

The Daniell cell is not used a great deal in practice, but its e.m.f. is practically constant at 1·09 V so that it may be used as a standard source of e.m.f. in the laboratory for measurement purposes. Its other advantages are that even for currents of 2 A or 3 A for long periods there is no polarisation, and that it is cheap to run.

Since the electrolytes diffuse through the porous pot, the cell must not stand open-circuited for long periods, and it must be thoroughly cleaned after use.

The Leclanché cell—wet type

The wet type of Leclanché cell is shown in Fig. 6.3. It consists of a glass jar containing a saturated solution of sal-ammoniac (ammonium chloride) with an amalgamated zinc rod and a carbon rod as the negative and positive poles respectively. A mixture of powdered carbon and manganese dioxide surrounds the carbon rod in a porous pot. There is a vent in the pitch at the top of the porous pot to allow the ammonia gas to escape. The electrolyte has a tendency to creep over the top of the glass jar, and the ammonium chloride to crystallise out. This is prevented by coating the upper part of the glass jar with pitch paint.

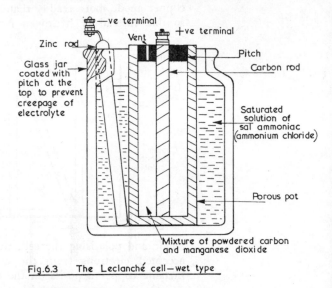

Fig.6.3   The Leclanché cell—wet type

When the external circuit is open, little chemical action takes place since amalgamation of the zinc rod with mercury stops local action.

When the external circuit is closed, the zinc reacts with the ammonium chloride to form zinc chloride, ammonia and hydrogen. Bubbles of hydrogen collect around the carbon and polarise the cell. The manganese dioxide oxidises the hydrogen to water, but at a slower rate than hydrogen is produced, so that the cell's e.m.f. gradually falls.

The ionic action of the cell is as follows. Ammonium chloride in solution dissociates into positive ammonium ions and negative chloride ions. When the external circuit is closed positive zinc ions enter the solution, releasing electrons via the zinc cathode to the external circuit. At the same time the positive ammonium ions accept electrons from the external circuit via the carbon anode, and become the electrically neutral ammonium radicals which immediately decompose to form hydrogen and ammonia gases. The ammonia dissolves in the water of the solution, and the hydrogen is changed to water by the depolariser, manganese dioxide. If a large current is taken from the cell for a long period of time, the rate of production of ammonia and hydrogen is greater than the rate at which the ammonia dissolves and the hydrogen is removed. The ammonia gas then escapes through the vent and the hydrogen polarises the cell.

When the external circuit is opened, the manganese dioxide continues to act until all the hydrogen is removed from the carbon.

Thus the Leclanché cell is most useful for intermittent work; for example for bell circuits and telephones. The e.m.f. of a Leclanché cell is about 1·5 V and it has an internal resistance of about 1 ohm.

The Leclanché cell—dry type

The dry type of Leclanché cell is shown in Fig. 6.4. The positive pole is a carbon rod that is sometimes fitted with a metal cap for connection purposes. The negative pole is the zinc container, and the electrical contact is usually made at the base of the container. Surrounding the carbon rod is a mixture of manganese

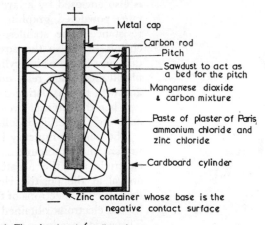

Fig.6.4 The Leclanché cell — dry type

dioxide and carbon. A wet paste of plaster of Paris, ammonium chloride and zinc chloride fills the space between the mixture and the zinc container. A cylinder of cardboard covers the zinc container and the top is covered with a sealing material such as pitch which in the larger cells has a vent provided to allow the ammonia to escape.

The action is similar to that of the wet type of cell.

The e.m.f. for a new cell is about 1·5 V. This quickly falls to 1·4 V which value is maintained until the cell is nearly exhausted. The cell is discarded when the cell's e.m.f. is about 1 V. The cell has a low internal resistance of the order of 0·1 to 0·3 ohms.

The advantages of this type of cell are its compactness, its portability and, if it is suitably sealed, its freedom from 'creepage' of the sal-ammoniac. If it is not suitably sealed care must be taken to prevent it from corroding the equipment with which it is used.

Mercury cell

A typical mercury cell is shown in cross-section in Fig. 6.5.

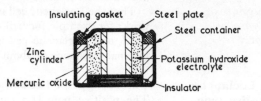

Fig.6.5 A mercury cell

Its negative electrode is a hollow cylinder of zinc foil or compressed zinc powder. This cylinder is surrounded by a layer of electrolyte which is a concentrated solution of potassium hydroxide and zinc oxide. The electrolyte is also enclosed by a layer of mercuric oxide that contains powdered graphite to reduce the cell's internal resistance. The stainless-steel container for all these chemicals is also the positive electrode. An insulating disc supports the zinc cylinder and electrolyte. A nickel-plated steel plate rests on the top of the zinc cylinder and with the insulating gasket seals the cell.

When the external circuit to the cell is closed the ionic action of the cell is as follows.

The positive mercury ions and negative oxygen ions present in the mercuric oxide travel towards the positive electrode and the negative electrode respectively. On arrival at the positive electrode the positive charge or deficiency of electrons of the mercury ions is neutralised by the electrons obtained from the positive electrode. This action produces mercury on the inside of the steel container. The negative oxygen ions meanwhile travel towards the negative zinc cylinder but combine with potassium ions and water in the electrolyte to re-form potassium hydroxide.

In the electrolyte the potassium hydroxide ionises to provide positive potassium ions which travel towards the mercuric oxide and negative hydroxide ions that move towards the negative zinc cylinder. These hydroxide ions give up their surplus negative electrons to the external circuit via the zinc. The hydroxide radical then combines with the zinc to form zinc oxide and water.

Since no gases are produced under normal conditions at either electrode by its chemical reactions there is no polarisation in a mercury cell. It therefore maintains virtually a constant terminal voltage of about 1·2 V to 1·3 V. Also it can be stored without loss of capacity for a long time at normal temperatures since there is no polarisation in this type of cell.

The constant e.m.f., the long storage life and its high output with a low mass make a mercury cell particularly suitable for use in miniature electronic circuits found in medical electronics, hearing aids, etc.

Production of e.m.f. by secondary cells

In a secondary cell the chemical reaction can be reversed by adding electrical energy to the cell. The process of supplying electrical energy to a secondary cell, where it is stored in the form of chemical energy in the chemicals of the cell, is known as charging. In the reverse process of taking electrical energy from the cell—that is, in the discharging process—the chemical reaction releases chemical energy for conversion to electrical energy when the load circuit external to the cell is closed. This reversibility of the chemical reaction in a secondary cell is the fundamental difference between a primary and a secondary cell. The types of secondary cell described in this chapter are the lead-acid cell and the nickel-iron and nickel-cadmium alkaline cells. Since secondary cells can accumulate electrical energy in the form of chemical energy, they are often referred to as *accumulators*.

The lead-acid secondary cell
The lead-acid secondary cell has two electrodes, each comprising a set of lead plates covered with lead compounds. The negative electrode has one more plate than the positive electrode, and the sets of plates are interleaved so that both sides of a positive plate face a negative plate. Such an arrangement prevents the positive plates from 'buckling' under heavy load currents. Separators of ebonite, celluloid or specially treated wood are placed between the plates to prevent them from touching. This avoids a short circuit from occurring between the positive and negative electrodes. The electrodes and separators are placed in a container of glass or any other suitable acid-resistant container. To complete a lead-acid cell, dilute sulphuric acid is added to the container until the plates are covered.

In the open-type cell shown in Fig. 6.6 the positive plates are usually formed and the negative plates *pasted*. The formed type of plate is made of lead plate by the long and expensive process of repeated charging and discharging. To make the pasted negative plates, a paste of sulphuric acid and red lead is pressed into a lead grid. Pasted plates are normally used for both electrodes in small cells where weight is a deciding factor, since pasted plates are less than half as heavy as formed plates with the same ampere-hour capacity.

Plates of the same polarity are fixed together on a common bar to form an electrode. The plates have lugs on them which with the common bar support the electrode, keeping them clear of the bottom of the container. The clearance space at the bottom is needed so that any material falling from the plates has room to collect without any danger of short-circuiting the electrodes.

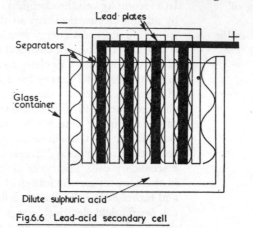

Fig.6.6 Lead-acid secondary cell

In discharging a lead-acid cell, the chocolate-brown lead peroxide on the positive plates and the grey porous lead of the negative plates in a fully charged cell slowly change into grey lead sulphate, releasing hydrogen from the acid at the same time. The hydrogen combines with the oxygen of the lead peroxide to form water which slowly reduces the specific gravity of the sulphuric acid to about 1·17 when the cell is fully discharged. A hydrometer, a device for measuring specific gravity, can be used to determine the state of the cell. The specific gravity of the acid in a fully charged cell is about 1·21.

In charging a lead-acid cell, the chemical reactions reverse: lead peroxide and porous lead re-form on the positive and negative plates respectively. The sulphate ions released from the lead sulphate of both positive and negative plates combine with the hydrogen in the water to form sulphuric acid. A corresponding quantity of oxygen from the water is absorbed, and forms brown lead peroxide on the positive plates. The colour of the positive plates—chocolate brown when fully charged to grey when discharged—gives a guide to the condition of the cell.

The e.m.f. of a cell—that is, its open-circuit terminal voltage—is about 2·4 V when fully charged and 1·85 V when discharged. A suitable d.c. voltmeter may be used to measure a cell's terminal voltage and thus to check its state. The e.m.f. of the cell falls quickly on discharge to 2 V, where it is maintained for a long period until

94

the cell becomes fully discharged and then rapidly falls to 1·85 V. It is not advisable to attempt to take load from a cell when it is in this condition.

The internal resistance of a secondary cell is usually low, of the order of 0·04 Ω, due to the large area of the positive and negative plates and their proximity to each other. Since the internal resistance is low, it is inadvisable to short-circuit the terminals of a secondary cell because of the high current that would result.

A summary of the changes that take place during charge and discharge in a lead-acid secondary cell is given below:

	Fully charged	Discharged
Positive plate:	brown-lead peroxide	grey-lead sulphate
Electrolyte:	dilute sulphuric acid	water
Specific gravity:	1·21	1·17
Negative plate:	grey-lead porous	grey-lead sulphate
E.M.F.:	2·4 V	1·85 V

Average p.d. per cell while discharging is 2 V

The alkaline secondary cells

The two main types of alkaline accumulators are made of nickel-iron and those made of nickel-cadmium.

The positive plates of both types contain nickel hydroxide held in finely perforated steel compartments set in rigid steel frames. Flakes of pure nickel or graphite are added to the nickel hydroxide to reduce its electrical resistance. In both cases a solution of potassium hydroxide with a specific gravity of approximately 1·17 is used as the electrolyte. The value of the specific gravity remains constant since the electrolyte does not enter the chemical reactions during the charge or discharge of an alkaline cell. For this reason a hydrometer is not used to check the state of charge of an alkaline cell.

The negative plates in a nickel-iron cell are made of iron oxide mixed with some mercuric oxide and enclosed in perforated steel plates. The mercuric oxide is added to reduce the plates' electrical resistance. In the nickel-cadmium cell, the active chemical enclosed in the perforated steel plates is cadmium, which is mixed with some iron to prevent the cadmium from losing its porosity and caking.

Ebonite rods separate the positive and negative plates. Steel spacing washers hold apart plates of the same polarity which are firmly bolted together on a steel terminal bolt. The whole assembly is placed in steel containers, and the electrolyte added to complete the cell. Wooden crates house the steel containers so that the cells are insulated from one another when they are connected together to form a battery.

A summary of the changes that occur in alkaline cells during charge and discharge is given below:

	Fully charged	Discharged
Positive plate:	higher hydroxide of nickel	lower hydroxide of nickel
Electrolyte:	potassium hydroxide unchanged	
Specific gravity:	constant at about 1·17	
Negative plate:	cadmium or iron	cadmium hydroxide or iron hydroxide
E.M.F./cell:	1·6 to 1·7 V	1·1 V

Average discharge p.d. across a cell is 1·2 V

Alkaline cells are robust mechanically and may be left in any state of charge without damage since they do not 'sulphate'. However, an alkaline cell costs more than a lead-acid cell of similar capacity, and in addition the average discharge p.d. per cell is 1·2 V compared with 2 V for a lead-acid cell. This means that a battery with a greater number of alkaline cells would be required to give the same p.d. as a battery of lead-acid cells, and the alkaline battery would cost more than an equivalent lead-acid battery.

Importance of a low internal resistance

It is important for a cell to have a low internal resistance in order for the internal voltage drop and the internal power loss to be kept as low as possible when the cell discharges.

Consider the circuit shown in Fig. 6.7. The voltage equation for the complete circuit is

$$E = V + Ir \text{ volts}$$

where E = e.m.f. of the cell or the open circuit terminal p.d. in volts

V = terminal p.d. of the cell or the voltage drop across the external load in volts

I = discharge current in amperes

r = internal resistance of the cell in ohms.

The internal voltage drop in the cell is Ir volts due to the current I amperes flowing through the internal resistance r ohms of the cell. Also, the power loss in the cell is I^2r watts. To keep both the internal voltage drop and the power loss down, the internal resistance of the cell must have a low value.

Fig.6.7

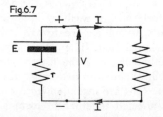

Methods of connecting cells to form a battery

If a greater d.c. voltage is required or more direct current is demanded by a load than can be provided by a single cell, then several cells have to be connected together to form a battery which will meet the load's requirements.

The three methods of connecting cells to form batries are:

1 the series method
2 the parallel method
3 the series-parallel method

The series method

In this arrangement the negative terminal of the first cell is connected to the positive terminal of the second cell, and the negative terminal of the second cell is connected to the positive terminal of the third cell. This procedure is repeated until all the cells have been connected in series. The positive terminal of the battery is the positive terminal of the first cell and the negative terminal of the battery is the negative terminal of the last cell.

Fig.6.8

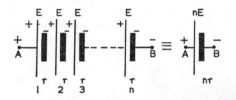

Fig.6.9

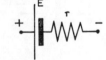

Fig.6.10

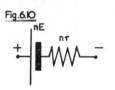

The resultant e.m.f. of the battery is the sum of the individual e.m.f.s of the cells. The internal resistance of the battery is the sum of the internal resistances of the cells. The current that flows through the external load is the same current that flows through each cell.

If a battery comprises n similar cells connected in series, each cell with an e.m.f. of E volts and an internal resistance of r ohms, then the e.m.f. of the battery is nE volts and its internal resistance is nr ohms (see Figs. 6.8, 6.9 and 6.10).

If in the series arrangement there is a reversed cell (see Fig. 6.11) this cell's e.m.f. opposes the sum of the e.m.f.s

Fig.6.11

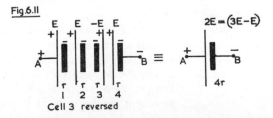

Cell 3 reversed

of the correctly connected cells, and the result is a reduction of the battery's e.m.f. The internal resistance is unchanged and equal to the sum of the internal resistances of all the cells.

The parallel method

In this arrangement, all the positive terminals of the similar cells are connected to a common terminal which forms the positive terminal of the battery, and all their negative terminals are connected together to form the battery's negative terminal.

For a battery of n similar cells in parallel, the e.m.f. of the battery is the same as the e.m.f. of a single cell —that is, E volts—but the battery's internal resistance is reduced to $\dfrac{r}{n}$ ohms, where r ohms is the internal resistance of each cell (see Fig. 6.12).

With the parallel connection, each cell contributes equally to the total current supplied to the external load. The total load current is the sum of the currents supplied by each parallel cell.

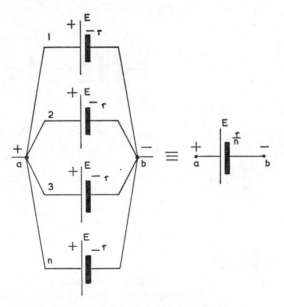

Fig.6.12 Battery of 'n' similar cells in parallel.

The series-parallel method

In this arrangement equal numbers of similar cells are connected in series, and the series groups are connected in parallel to form the battery. The e.m.f. of a battery comprising p groups of cells in parallel, with each group having n similar cells in series, equals the total e.m.f.

of one of the series groups of cells. Thus if each cell has an e.m.f. of E volts and an internal resistance of r ohms, the battery's e.m.f. is nE volts and its internal resistance is $\frac{nr}{p}$ ohms. The number of groups in parallel p does not affect the battery's e.m.f. but it reduces the internal resistance of the battery (see Fig. 6.13).

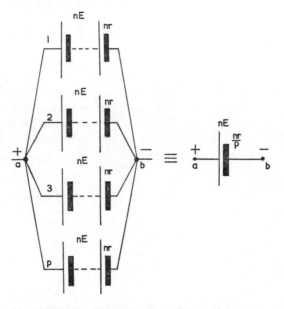

Fig.6.13 Battery of 'p' groups in parallel, each group comprising 'n' similar cells in series.

Production of an e.m.f. by electromagnetic induction

An electromotive force is produced in a conductor when the conductor cuts or is cut by a magnetic field of flux. The e.m.f. is generated only while there is relative motion between the magnetic flux and the conductor.

This method of generating or inducing an e.m.f. is known as electromagnetic induction. Faraday discovered this effect and that the magnitude of the induced e.m.f. is proportional to the rate at which the magnetic flux is cut.

Faraday's law of electromagnetic induction

This law states that when an e.m.f. is generated in a conductor by magnetic flux cutting the conductor, the magnitude of the induced e.m.f. is proportional to the rate at which the magnetic flux cuts the conductor.

In symbols, $E \propto \dfrac{\Phi}{t}$

where E = the average e.m.f. generated in the conductor
Φ = the magnetic flux cut by the conductor
t = the time taken to cut the flux

In the International System of Units (SI) the magnitude of the e.m.f. in volts is given by $E = \dfrac{\Phi}{t}$ volts $\qquad$ **(6.1)**

where E = the average e.m.f. induced in the conductor in volts

if $\quad \Phi$ = the magnetic flux cut by the conductor in webers and

$\quad t$ = the time taken to cut the magnetic flux in seconds

This version of Faraday's law is sometimes called the *flux cutting rule.*

The angle at which the magnetic flux is cut is important since maximum e.m.f. is induced when the relative motion of the conductor is at right angles to the magnetic flux; that is, when the magnetic flux is cut at right angles. No e.m.f. is generated in the conductor if the conductor travels parallel to the magnetic flux (see Fig. 6.14).

Fig.6.14

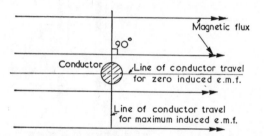

When a conductor CD travels at right angles to the magnetic flux at a constant velocity as shown in Fig. 6.15 the magnitude of the induced e.m.f. in the conductor is given in the SI by the expression:

$$E = Blv \text{ volts} \qquad \textbf{(6.2)}$$

where E = the e.m.f. in volts induced in the conductor

$\quad B$ = the magnetic flux density in teslas of the uniform magnetic field

$\quad l$ = the active length of the conductor in metres, i.e. the length of conductor cutting the flux

$\quad v$ = the velocity in metres/second of the conductor at right angles to the magnetic field.

Since motion of the conductor is involved in the production of this e.m.f. it is sometimes referred to as the dynamically induced e.m.f. This method is used in electromagnetic machines that convert mechanical energy to electrical energy—that is, generators, and so it is also referred to as the generator principle.

In order to detect the induced e.m.f. a galvanometer is connected to the ends of the conductor. A galvanometer is a sensitive d.c. instrument, usually with a

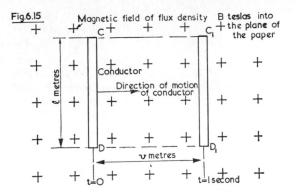

Fig.6.15

Magnetic field of flux density B **teslas** into the plane of the paper

Conductor

Direction of motion of conductor

ℓ metres

v metres

$t=0$

$t=1$ second

centre zero scale. It gives full-scale deflection when a direct current of the order of milliamperes passes through it due to a p.d. of the order of millivolts applied to its terminals.

The direction of the deflection of the galvanometer needle depends on the direction of the current through the instrument, and, therefore, on the polarity of the p.d. applied to its terminals.

With the galvanometer terminals connected to C and D, the conductor forms part of a complete circuit. Thus the flux cut by the conductor CD may be regarded as the change in flux linking the circuit. Also, the circuit may be regarded as a single turn loop or coil and the change of flux linking the circuit as the number of interlinkages between the electric and the magnetic circuits. If CD represents *one* side of a coil with N turns and the change of flux is Φ webers in t seconds then N is the change of flux linkages in t seconds.

An alternative statement of Faraday's law of electromagnetic flux linking the circuit, the magnitude of the induced e.m.f. is proportional to the rate of change of flux-linkages with the circuit.

In symbols, $E \propto \dfrac{\Phi N}{t}$

$$\text{In SI units } E \propto \dfrac{\Phi N}{t} \qquad (6.3)$$

where $E=$ the e.m.f. induced in volts
when $\Phi N=$ the change in flux-linkages in weber-turns
and $t=$ the time taken in seconds for the change in flux linkages.

Definitions of a volt

The unit of e.m.f.—a volt—may be defined from the various versions of Faraday's law of electromagnetic induction as follows:

1 The volt is the unit of e.m.f. induced in each metre of a conductor that is travelling at right angles to a magnetic field of flux density 1 tesla with a velocity of 1 m/s.

$\therefore E = Blv$ volts   and if $B = 1$ T   then $E = 1$ V
$$l = 1 \text{ m}$$
$$v = 1 \text{ m/s}$$

2 The volt is the unit of e.m.f. induced in a single-turn circuit when the rate at which the flux linking the circuit is changing or the rate at which the flux is cutting the circuit is 1 Wb/s.

$$\therefore E = \frac{\Phi N}{t} \quad \text{if } N = 1 \quad \text{then } E = \frac{\Phi}{t} \text{ volts}$$

$$\text{and if } \frac{\Phi}{t} = 1 \text{ Wb/s} \quad \text{then } E = 1 \text{ V}$$

3 The volt is the unit of e.m.f. induced in a circuit when the flux-linkages with the circuit are changing at the rate of 1 weber-turn/second.

$$\therefore E = \frac{\Phi N}{t} \text{ volts} \quad \text{if } \frac{\Phi N}{t} = 1 \text{ weber-turn/second}$$

$$\text{then } E = 1 \text{ V.}$$

Methods of inducing e.m.f. and current

Relative motion between a circuit and a magnetic field or change of flux linkages between electric and magnetic circuits will produce an induced e.m.f. in the circuit. This e.m.f. then sends a current through the circuit if the circuit is complete.

These e.m.f.s and currents can be produced:
1 by the rotation of a coil in a stationary field
2 by the movement of a permanent magnet towards or away from a stationary coil
3 by the movement of a solenoid—that is, a current-carrying coil or electromagnet—towards or away from a stationary coil
4 by the variation of current in one stationary coil while it is near a secondary stationary coil.

Since method **1** depends on physical movement of the coil in which the e.m.f. is induced, the e.m.f. generated by this means is referred to as *dynamically induced e.m.f.*

In methods **2**, **3**, and **4** the coils in which the e.m.f. is generated are stationary, and the e.m.f. induced by these methods is therefore known as statically induced e.m.f.

It must be noted that there is no difference in the nature of the e.m.f.s produced by any of these methods.

In method **4** the flux linkages with the second coil are changed by altering the magnitude of the current supplied to the first coil.

Method **4** is the method used in the double-wound

transformer to produce an e.m.f. in the secondary winding when an alternating current is supplied to the primary winding.

Direction of induced e.m.f. and current

The expressions 6.1, 6.2 and 6.3 give the magnitude of induced e.m.f. However, the direction of the e.m.f. must also be known in order to predict the direction in which the induced current will flow under the influence of the induced e.m.f. when the conductor or coil forms part of a complete circuit.

Lenz's law

This law predicts the direction of the induced current and e.m.f. It may be stated as follows:

'An e.m.f. is induced in such a direction that when the circuit is complete, the induced current will flow so as to oppose the cause producing the e.m.f.'

In Fig. 6.16 the galvanometer completes the circuit so that an induced current will flow in the circuit under the influence of the induced e.m.f. in conductor CD.

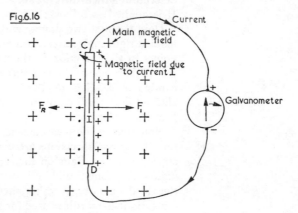

Fig.6.16

The cause producing the induced e.m.f. of magnitude $E = Blv$ volts in the conductor is the force F_1 which moves the conductor from left to right at right angles to the stationary field of flux density B teslas. By Lenz's law the induced current I will flow so as to oppose this force F_1—in other words from D to C in the conductor. In this direction the current sets up a magnetic field that strengthens the main field on the right-hand side of the conductor and weakens the main field on the left-hand side of the conductor. The force due to the interaction of these fields is given by $F_R = BIl$ newtons. This force F_R opposes the main motive force F_R and tends to move the conductor CD from right to left— that is, in a direction opposite to the actual direction of motion of the conductor.

Conventionally, end C of the conductor is regarded as positive with respect to end D since the induced current flows from C through the external circuit to D. The conductor CD is actually a source of e.m.f. due to electromagnetic induction. Fig. 6.17 shows the conventions for a similar circuit using a chemical source of e.m.f. such as a simple cell to give the same direction of deflection of the galvanometer needle.

Fig. 6.17

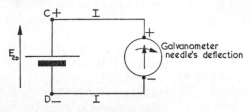

Fleming's right-hand rule

Although the direction of the induced e.m.f. and current can be deduced by using Lenz's law, a quick method for determining their direction is Fleming's right-hand rule. It is applied as follows:

Hold the first two fingers and the thumb of the right hand at right angles to each other. Then if the First finger points in the direction of the main Field or Flux, and the thuMb points in the direction of Motion of the conductor relative to the magnetic field, then the sECond finger will point in the direction of the induced E.m.f. and Current.

Note that the bold capitals serve as memory aids, and that if any two directions are correctly indicated by the appropriate two digits on the right hand, then the third digit will automatically indicate the correct direction for the third quantity. Another memory aid is that Fleming's Right-hand rule is used for the geneRator principle.

Fig. 6.18

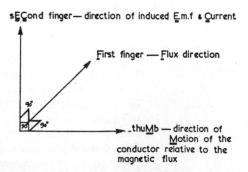

Expressions which give the magnitude and direction of the induced e.m.f. are:

$$E = \frac{-\Phi}{t} \quad \text{volts} \qquad \textbf{(6.4)}$$

$$E = -Blv \quad \text{volts} \qquad \textbf{(6.5)}$$

$$E = \frac{-\Phi N}{t} \quad \text{volts} \qquad \textbf{(6.6)}$$

The minus signs in the expressions just given indicate that the direction of the induced e.m.f. is in accordance with Lenz's law.

Simple types of a.c. and d.c. generators

Simple alternator

In the simple alternating current generator or alternator shown in Fig. 6.19 the armature winding is a single-turn coil *abcd* that is rotated by a prime mover in an anti-clockwise direction at a speed of n_s revolutions per second in a stationary magnetic field of flux density B

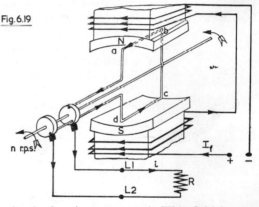

Fig. 6.19

teslas (webers/square metre). This field is provided by a stationary field system of a two-pole electromagnet. The ends *a* and *d* of the armature winding are connected to two brass slip rings. Two carbon brushes resting on them make the electrical connection from the ends *a* and *d* to the output terminals L1 and L2 respectively of the alternator. The external load is connected across these terminals.

Fleming's right-hand rule for a generator when applied to the above alternator shows that it generates e.m.f. in a coil side in the positive or forward direction when the coil side moves anti-clockwise under a north pole, and in the negative or reverse direction when the coil side moves anti-clockwise in relation to the magnetic flux entering a south pole. Therefore at the instant shown in Fig. 6.19 terminal L1 is positive with respect to terminal L2 and the current is then leaving the alternator at terminal L1, passing through the external load and entering at terminal L2 at that instant. The e.m.f. generated at that instant is maximum in the positive conventional sense since the coil sides *ab* and *dc* are a pole pitch apart: they travel at the same velocity through the flux of opposite poles and maximum e.m.f. is induced in each of them in opposite directions. These e.m.f.s per coil are connected in series aiding by the back connection *cb* of the coil.

For every revolution of the armature in a two-pole generator there is one reversal of the direction of the induced e.m.f. since coil side *ab* travels for half a revolution under a north pole and for the other half under a south pole. This means that terminal L1, which is always connected to the start of the coil or end *a* via the same brush and slip ring, reverses once in polarity each time coil *ab* and *cd* pass a pair of poles. When a purely resistive load is connected to terminals L1 and L2 the current that flows through the complete circuit due to the influence of the e.m.f. also changes direction once for each revolution of the two-pole alternator's rotor, or once for each cycle of the alternating e.m.f. and current when its load circuit is closed.

The variation with time of the e.m.f. and p.d. of L1 with respect to L2 is shown in Fig. 6.20. For a resistive load this graph also represents the alternating current's variation against time to a different scale.

Fig.6.20

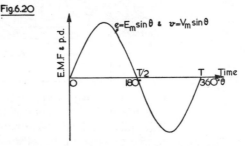

The number of reversals of polarity per second is known as the electrical frequency—f hertz—of the alternating e.m.f. and current. The relationship between the frequency of the electrical quantities, the speed of rotation and the number of pairs of poles on the alternator is

$$f = n_s p \text{ hertz} \tag{6.7}$$

where f = frequency of the alternating e.m.f. and current in hertz.

p = number of pairs of poles on the alternator

n_s = speed or rotation of the alternator or the driving speed of the prime-mover in revolutions/second. This speed is also called the synchronous speed of the alternator.

The simple d.c. generator

The difference between the simple d.c. generator shown in Fig. 6.21 and the simple alternator of Fig. 6.19 lies in the different means used to connect the internal rotating armature winding to the external stationary load circuit. In the alternator the connection is made by means of two slip rings, whilst in the simpler two-pole d.c. generator the connection is made via a two-part commutator.

This commutator as shown in Fig. 6.21 always connects the coil side that is travelling anti-clockwise under the north pole to brush B1. It is from this brush that current always leaves the generator, and therefore brush B1 is always positive with respect to brush B2. Consequently terminal L1 is also always positive with respect to terminal L2, since current conveniently leaves the source and enters a load at their positive terminals.

Fig.6.21

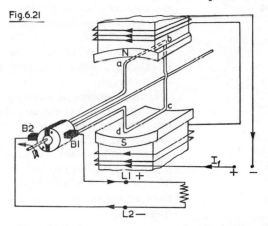

Fig. 6.22 shows a sequence of coil positions and the direction of the induced e.m.f. in these positions during

Fig.6.22

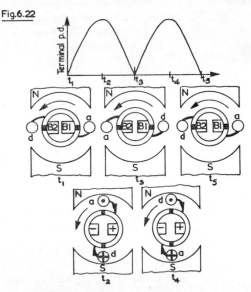

one complete revolution of the armature winding at five instants in time. The commutator automatically reverses the negative half-cycle of the alternating e.m.f. induced or generated between the ends a and d; that is, between the start and finish of the coil as far as the external circuit is concerned. The reversal of the e.m.f. and current in the coil has to occur while the commutator segments are short-circuited by the brushes. This process is known as commutation. It is the function of the commutator and brush gear to perform this process safely.

Alternating current

When a steady electric current flows along a conductor, the number of electrons that drift in one direction axially through any cross-section in each second is a fixed number. Therefore the total electric charge in coulombs that crosses the section in one direction in each second is constant. Such a current of electricity is called a *direct current*.

If the numbers of electrons that cross a section vary from instant to instant and their axial drift across the conductor section is not always in the same direction, the electric current is known as an *alternating current*. Consequently an alternating current is one that changes periodically with time both in magnitude and direction. The generation of an alternating e.m.f. and current has been discussed in the section on the simple alternator (page 106).

Wave form or wave shape

A graph that shows the variation in magnitude and direction of an alternating quantity over a series of consecutive instants in time is known as its wave form or wave shape. Two examples of alternating current wave forms are shown in Figs. 6.23a and b.

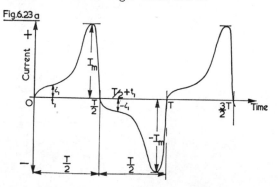

Fig.6.23 a

The circuit diagram in Fig. 6.24a shows direct current I_{ab} flowing in the conventional forward or positive direction through an external load resistor R. The d.c.

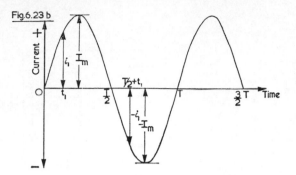

Fig.6.23 b

source of a battery of cells, connected as shown, ensures that terminal a always possesses a positive potential with respect to terminal b, and that a steady current I_{ab} flows into the load resistor at terminal a and out at terminal b. Fig. 6.24b shows the graph of the direct current I_{ab} against time.

In Fig. 6.23c an a.c. source supplies an alternating current to the load resistor, since the potential and the polarity of terminal a with respect to terminal b changes periodically with time. When terminal a has a positive potential with respect to terminal b, say at instant t_1

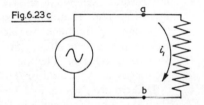

Fig.6.23 c

seconds, the instantaneous current i_1 flows in the forward or positive direction from terminal a through the load to terminal b. The height i_1 above the horizontal time axis at instant t_1 in Fig. 6.23b represents to a suitable scale the magnitude of the instantaneous current. The fact that it is above the time axis indicates that the current flows in the conventional positive or forward direction at the particular instant in time under consideration; namely at instant t_1 seconds. Similarly at instant t_2 when the terminal a has a negative potential with respect to terminal b, the current flows from terminal b through the load to terminal a in the reverse or negative direction. The vertical distance i_2 of the current wave below the point on the time scale corresponding to instant t_2 seconds represents this reverse current in magnitude and direction. Thus the ordinates of a wave form give the magnitude and direction of an alternating quantity at any instant in time.

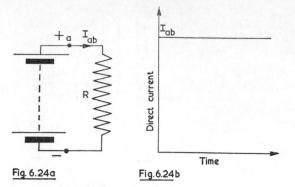

Fig. 6.24a   Fig.6.24b

Periodic time or period (T seconds)

One complete reversal or alteration of the alternating quantity is called its cycle. The time in seconds taken to complete a cycle is known as the periodic time, or period of the alternating quantity, and is represented by the symbol T.

The values of the alternating quantity at instants in time that differ by the periodic time T are identical, so that the wave form repeats itself every T seconds.

Frequency (f hertz)

The number of cycles of the wave form that occur in one second is known as the frequency of the alternating quantity. The SI unit of frequency is the hertz (Hz). The standard frequency for electrical alternating quantities in power generation and supply systems in Great Britain is 50 Hz. The period corresponding to this frequency is $\frac{1}{50}$ second or 0·02 second.

If there are f cycles in 1 second
then the time for 1 cycle is $1/f$ second,
thus the period T is $1/f$ second

$$\therefore T = 1/f \text{ second} \tag{6.8}$$

and $\therefore f = 1/T$ cycles/second or hertz   (6.9)

The time for a complete cycle or 1 period is also referred to as 360 degrees electrical.

Thus $T = 1/f$ second $= 260$ degrees electrical (6.10)

This is a very important time measure in electrical engineering.

Peak or maximum value of an alternating quantity

The part of a cycle that represents the forward or positive direction of an alternating quantity is known as its positive half-cycle, and that representing the reverse or negative direction is its negative half-cycle. If the negative half-cycle is a replica of the positive half-cycle but with negative ordinates at instants in time that differ by half a period from the corresponding positive ordinates, the wave form is called a symmetrical wave form. Figs. 6.23a and 6.23b are examples of symmetrical

110

wave forms, since the positive current i_1 at instant t_1 has the same magnitude as $-i_1$ at instant $(t+T/2)$.

The peak or maximum value of an alternating quantity is the greatest value it attains in any half-cycle. For symmetrical wave forms the positive and negative peak values are equal in magnitude but opposite in direction. In Figs. 6.23a and b the positive and negative peak currents are equal in magnitude but flow in opposite directions through the circuit.

Sinusoidal wave forms

The equation of the alternating current wave form shown in Fig. 6.23b is as follows:

$$i = I_m \text{ sine } \theta \tag{6.11}$$

where i = value of the alternating current in amperes
at instant t

I_m = maximum value of the alternating current in
amperes

θ = angle that corresponds to instant t in the
cycle. (Symbol θ is small Greek letter theta)

If is measured in degrees, then $\dfrac{\theta}{360} = \dfrac{t}{T}$

If is measured in radians, then $\dfrac{\theta}{2\pi} = \dfrac{t}{T}$

Then $i = I_m \text{ sine } \theta$ may be written as:

$$i = I_m \text{ sine } (360 \text{ ft}) \text{ where } \theta = 360 \text{ ft degrees} \tag{6.12}$$

or $i = I_m \text{ sine } (2 \text{ ft}) \text{ where } \theta = 2\pi \text{ ft radians} \tag{6.13}$

All the above equations just given for instantaneous current in amperes show that the current wave form of Fig. 6.23b obeys a sine law. Consequently it is described as a sinusoidal wave form, as are all wave forms that obey a sine law. The sine law is a very important wave form in electrical engineering. It is the standard wave form used in electrical engineering to represent alternating currents and voltages because of its mathematical simplicity and versatility. It is the ideal wave form for electrical power generation, transmission and utilisation. It lends itself to phasor (formerly called vector) representation.

Wave forms that do not obey a simple sine law are known as non-sinusoidal wave forms. Fig. 6.23a shows a wave form that is obviously not a sine wave. It is therefore described as a non-sinusoidal wave form although it is symmetrical.

Average value of an alternating quantity

The average value of an alternating quantity is the average or mean height of its wave form.

Consider the wave form shown in Fig. 6.25. This is a non-sinusoidal symmetrical wave form of current. Since it is symmetrical the average value of the alternating current (I_{av}) for the complete cycle is zero. Over either the positive or negative half-cycle the average current is:

$$I_{av} = \frac{i_1 + i_2 + i_3 + \dots + i_n}{n} \text{ amperes} \qquad (6.14)$$

where i_1, i_2, i_3. ... i_n are n equally spaced mid-ordinates of current.

The number of mid-ordinates taken in each half-cycle is usually six; that is, $n = 6$. Then

$$I_{av} = \frac{i_1 + i_2 + i_3 + i_4 + i_5 + i_6}{6} \text{ amperes} \qquad (6.15)$$

Fig.6.25

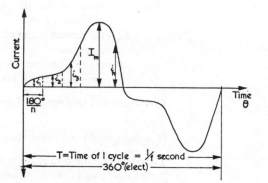

For a pure sine wave of current, over a half cycle:

$$I_{av} = \frac{2}{\pi} I_m = 0 \cdot 637 \, I_m \qquad (6.16)$$

over a complete cycle:

$$I_{av} = 0 \qquad (6.17)$$

Similar relationships hold for the average value of an alternating voltage:

Over either a positive or negative half-cycle:

$$V_{av} = \frac{v_1 + v_2 + v_3 + \dots + v_n}{n} \text{ volts} \qquad (6.18)$$

where v_1, v_2, v_3 .... v_n are n equally spaced mid-ordinates of voltage. The number of mid-ordinates taken in each half-cycle is normally six.
Thus

$$V_{av} = \frac{v_1 + v_2 + v_3 + v_4 + v_5 + v_6}{6} \text{ volts} \qquad (6.19)$$

For a pure sine wave of voltage over a half-cycle:

$$V_{av} = \frac{2}{\pi} V_m = 0 \cdot 637 \, V_m \qquad (6.20)$$

over a complete cycle:

$$V_{av} = 0 \qquad (6.21)$$

An alternative interpretation of the average value of an alternating quantity for half a cycle is to regard it as the height of a rectangle that has the same base and an area equal to that enclosed by half the wave form and the time axis.

This method is useful when the area enclosed by the wave form and the time axis is a simple geometrical figure whose area can be easily found. For more difficult geometrical shapes the mid-ordinates rule is used.

Root mean square value of an alternating current ($I_{r.m.s.}$)

The root mean square (r.m.s.) value of an alternating current $I_{r.m.s.}$ is the value of a direct current that produces the same heating effect as the alternating current when both currents flow independently through equal electrical resistances for the same interval of time.

The average heating effect of an alternating current with a wave form as shown in Fig. 6.26 is obtained by applying the mid-ordinate rule to n equally spaced ordinates of current on the graph of the variation in heating effect over a half-cycle.

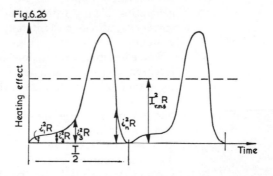

Fig.6.26

$I_{r.m.s.}$ = square *root* of *mean* of the *squares* of the instantaneous values of current
$I_{r.m.s.}$ = root mean square value of the alternating current
$I_{r.m.s.}$ = r.m.s. value of the alternating current.

$$I_{r.m.s.} = \sqrt{\left(\frac{i^2_1 + i^2_2 + i^2_3 + \dots + i^2_n}{n}\right)} \qquad (6.22)$$

The number of mid-ordinates taken is usually six; that is, $n = 6$. Then

$$I_{r.m.s.} = \sqrt{\left(\frac{i^2_1 + i^2_2 + i^2_3 + i^2_4 + i^2_5 + i^2_6}{6}\right)} \qquad (6.23)$$

For a sinusoidal alternating current

$$I_{r.m.s.} = \frac{1}{\sqrt{2}} I_m = 0.707 \, I_m \qquad (6.24)$$

Since the p.d. across a resistor is directly proportional to the current flowing through it at any instant, similar expressions apply to the root mean square value of an alternating voltage. The expressions are:

113

$$V_{\text{r.m.s.}} = \sqrt{\left(\frac{v^2{}_1 + v^2{}_2 + v^2{}_3 + \ldots + v_n{}^2}{n}\right)} \qquad \textbf{(6.25)}$$

$$V_{\text{r.m.s.}} = \sqrt{\left(\frac{v^2{}_1 + v^2{}_2 + v^2{}_3 + v^2{}_4 + v^2{}_5 + v^2{}_6}{6}\right)} \qquad \textbf{(6.26)}$$

For a sinusoidal alternating voltage:

$$V_{\text{r.m.s.}} = \frac{1}{\sqrt{2}} \; V_m = 0\cdot707 \; V_m \qquad \textbf{(6.27)}$$

A.C. and d.c. supplies

The electricity supply provided by the local supply authority is known as an alternating current (a.c.) supply, so called because its direction of current flow is continually changing. The number of changes per second, *or cycles*, is referred to as the frequency of the supply. In this country the frequency is 50 hertz (cycles per second) and in America 60 hertz (cycles per second).

There is another type of supply called a direct current (d.c.) supply. This may be obtained from private generating plants, from batteries, or from alternating current. A direct current only flows one way around a circuit.

For some purposes it does not matter if the supply is alternating or direct, but for others it is extremely important. Electric heaters will work with a.c. or d.c. supplies. This is also true of most lighting equipment but not of discharge lighting, where the control gear varies according to the type of supply. Direct current must be used for electro-plating and for battery charging. The small motors used in domestic appliances such as vacuum cleaners and hair dryers can often be used with either kind of supply, and are called universal motors. Larger motors are normally wound for one or the other.

The d.c. motor has the advantage that its speed can be altered by means of simple control equipment. This is most important, as many machines must be able to work with a wide range of speeds. For some applications such as electric railways the advantages are so great that large quantities of electrical energy are converted from alternating to direct current. This is done by means of devices known as *rectifiers*.

A.C. motors are for the most part single-speed machines. Variable-speed a.c. machines are complex in design, and therefore very expensive. In consequence their use is rather limited.

Alternating current, however, has one advantage which completely outweighs all those of direct current. Whereas the voltage of a d.c. supply can only be altered by means of rotating machinery, that of an a.c. supply can be stepped up or down as needed by means of transformers.

These have no moving parts, and so are relatively cheap and easy to make and operate.

In delivering electrical energy from one place to another, losses are bound to occur, and their extent depends upon the size of the current flowing. For a given power the larger the voltage used the smaller the current needed, and therefore the smaller the resulting losses. With fixed machinery, much higher voltages can be dealt with than with rotating machinery. It follows that a.c. transmission is the more economical than d.c. transmission.

In addition the transformer gives a.c. systems a degree of flexibility not possible with a direct-current source.

The principle of the transformer is discussed in a later section of this chapter.

Two-wire mains supply (d.c. or a.c.) one side earthed

A two-wire installation may be connected to an alternating current or to a direct current source of supply. This source of supply may be earthed at its origin or at some other point, or it may not. The design of a two-wire installation will be determined in some respects by whether the source feeding it is earthed or otherwise.

At this stage only a two-wire earthed supply is considered.

Two-wire earthed supply

When an installation is receiving its energy from an earthed supply, single-pole switches may be used to control its various sections. All such switches must be connected in the line, positive or negative outer conductor (Fig. 6.27). Under no circumstances can a single-pole

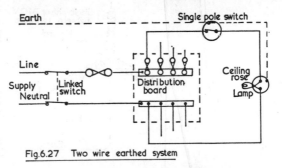

Fig.6.27 Two wire earthed system

switch be used in the neutral or other earthed conductor. If it were, the apparatus could appear to be 'dead' while in actual fact containing 'live' conductors. If for any reason it is decided to switch both line and neutral (or positive and negative), the switch used should be of a type that will break the two conductors at the same time. Such switches are known as double-pole linked switches. Circuit breakers should likewise be designed to break only the line conductors or both conductors simultaneously. Fuses and thermostats should be connected in the line or non-earthed conductor.

Voltage between consumers' terminals

Different consumers will have different reasons for requiring an electrical supply. Some will only need it for domestic purposes such as lighting or cooking, and for these people a 240 volt supply is sufficient. The 240 volts will be brought to them by a pair of conductors, one called the *line* and the other the *neutral*. This arrangement is called a single-phase supply, and the 240 volts mentioned is the nominal potential difference between conductors.

A consumer who has a heavier load will receive a 415/240 volt supply. In this case four wires are necessary —three lines and one neutral. The voltage between any two of the lines is a nominal 415 volts, and between any line and neutral it will be 240 volts (Fig. 6.28).

The area electricity boards are required by law to ensure that the pressure at the consumer's terminals is at no time 6% greater or less than the declared nominal voltage.

Voltage from mains to earth

As has already been mentioned, there must be a continuous path if an electric current is to flow. Usually the path for a current consists of wire or some other

Fig.6.28   Voltages between consumers terminals & earth.

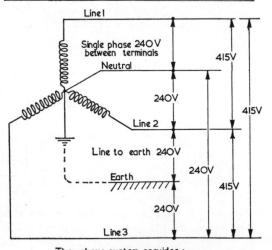

The above system provides :—
3 phase supply, 415V between lines.
1 phase supply , 240V line to neutral,
240V line to earth

current-carrying metal. In the case of a wet cell it will be a liquid, and in neon lighting it will be a gas. It must also be remembered that the mass of the earth can behave like a conductor. This is because it can absorb or give off very large numbers of electrons without showing any noticeable effect.

The neutral point in a public supply system is always connected to the earth both for protection and to assist in discovering faults. It follows that at any consumer's installation the earth will be present as an additional conductor, and that a voltage will be present between line and earth as well as between line and neutral. This potential between line and earth will in ordinary conditions be the same as the potential from line to neutral.

Simple a.c. systems

Single-phase two-wire system
This has been discussed in a previous section on page 115 of this chapter.

Three-phase three-wire system
Bulk supplies of electricity from the power stations are transmitted from place to place by means of three-wire systems operating at high voltages. One of these systems, known as the National Grid, operates at 132 000 volts (132 kV) between any pair of lines. Another system called the *Supergrid* operates at 275 000 volts (275 kV) and 400 000 volts (400 kV) between lines (see Fig. 6.29).

Fig.6.29   Three phase, 3 wire system.

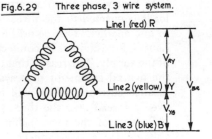

Mesh or delta connection of the windings of either a 3 ph alternator or 3 ph transformer secondary. Windings may also be star connected as shown below :—

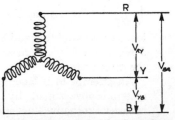

In both cases $V_{RY} = V_{YB} = V_{BR} = V_L$ = line voltage

Three-phase four-wire system

The high voltages of the three-phase transmission systems are stepped down at some convenient local substation, and the system becomes a four-wire arrangement. This is done by star-connecting the lower voltage side of the transformer, earthing the star point and running an additional wire from the star point (see Fig. 6.30).

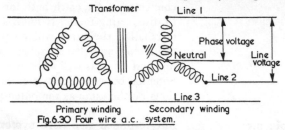

Fig. 6.30 Four wire a.c. system.

The wire connected to the common point of the three windings is called the neutral. The remaining wires are termed the lines. The four-wire system is used to supply domestic consumers at 240 V by taking a connection from any one line and the neutral. Small industrial consumers take the four wires with a voltage between lines of 415 V as well as 240 V from line to neutral. A.C. supplies obtained from only one line and the neutral are termed single-phase supplies.

Loading three-phase systems

Where a number of consumers are supplied from a 3 ph four-wire system an attempt is made to make the loads connected to the three lines approximately equal to one another. This could be done if there were sixty houses, for example, by connecting twenty to each phase of the supply. The reason that this is desirable is because the neutral conductor of a system only carries out-of-balance currents, which will be small if the loads are evenly spread over the three phases.

Because of the relationship between line and phase voltages for a star-connected system of three-phases, namely, line voltage is equal to $\sqrt{3}$ times the phase voltage ($V_L = \sqrt{3}Vph$), it is customary to have an r.m.s. line voltage of 415 V in order to obtain 240 V r.m.s. $\left(\dfrac{415}{\sqrt{3}}\right)$ between any line and neutral for single-phase loads such as lighting and heating and to supply balanced 3 ph, 415 V loads such as 3 ph induction motors from the three lines.

An attempt to balance the load taken from the 3 ph system is always made by sharing the single-phase loads between alternative lines and neutral; for example, single-phase load 1 may be connected across red line and neutral, single-phase load 2 across yellow line and neutral, single-phase 3 across blue line and neutral and for the next three 1 *ph* loads the process repeated. Under such conditions of loading, the three-phase system is seldom balanced since there is no guarantee that the single-phase loads are equal or that they are in use at the same instant even if they were equal. When such a case arises, the out-of-balance current returns from the load to the source through the neutral wire.

Earthing systems

An earthing system can be said to have two functions. The first of these is to ensure that no dangerous difference in potential can occur between metals in the region of an installation. The second is to provide a means of disconnecting the supply if a breakdown in insulation should occur, allowing current to flow to earth. Three different methods are used to obtain these results: solid earthing, the use of earth leakage circuit breakers and protective multiple earthing (PME).

Solid earthing

In solid earthing the consumer's earthing terminal is directly connected by means of an earthing lead to an electrode or electrodes in contact with the main mass of earth. This may be done by the supply authority providing a path back to the earthing point at the source of supply. In some cases this path will be by way of the armouring and sheath of an underground cable system. Alternatively the consumer may drive a suitable copper rod into the ground or bury a system of rods or tubes.

For this arrangement to be satisfactory, under fault conditions the impedance to earth must be very low— low enough to allow sufficient fault current to blow a protective fuse or trip an overcurrent circuit breaker. The fault current must exceed three times the rating of a fuse with a fusing factor over 1·5, or 2·4 times the rating of a fuse with a fusing factor of less than 1·5, or 1·5 times the tripping current of an overcurrent circuit breaker. The fusing factor of a fuse is the ratio of the current needed to melt it to its normal safe load current.

Earth-leakage circuit breakers

In some situations, for example in rocky regions, it is difficult to obtain an effective low-resistance earth path. If the impedance of an earth path is high the current that flows under fault conditions will be too small to blow a fuse or trip an overcurrent circuit breaker. In such a case an earth-leakage circuit breaker will be used.

These are circuit breakers which are designed to detect even very small earth leakage currents and to disconnect the supply when such currents flow.

Protective multiple earthing (PME)

Normally the supply authority is only allowed to earth the neutral of a system at one point. In some cases it is expedient to use the neutral of a system as an earth return path. There the earth continuity conductor of each installation is earthed and bonded to the neutral conductor at the intake. Special permission is needed for this, as laid down by the Electricity Regulations 1937.

Earthing an installation

Every installation must have an earth terminal placed close to the intake point. From this terminal, earth continuity conductors (e.c.c.) should be taken throughout the installation. All metal work in and around the installation, apart from current-carrying conductors, should be connected to an e.c.c. unless specifically exempted. Also jointed to the e.c.c. are the earthed terminals at the socket outlets, lighting points and switches in the installation. The core, metal case and one point on the secondary winding of each transformer must also be connected to the e.c.c. apart from the exemptional cases mentioned in regulation D8. The pipe-work of gas and water services and also structural steelwork must be connected to the e.c.c. after the main earthing system has been completed. This is known as bonding.

Earth continuity conductors (e.c.c.)

An e.c.c. is the link between any metal to be earthed and the consumer's earth terminal or earth leakage trip. The name implies a wire but in fact an e.c.c. may be made up of a number of different things. In some cases it will be a stranded or solid conductor forming part of a sheathed cable or cord. In other cases it may be a separate conductor, insulated or otherwise. Yet again it could be the sheath of a metal-covered cable or even part of a conduit or trunking system. In most cases, however, the e.c.c. is a combination of a number of these things. Pipes of other services and structural metalwork may not be used as part of an e.c.c.

When a composite earth continuity conductor is being used every care should be taken at joints to keep the total resistance to as low a value as possible. This will mean that paint and dirt must be removed where any metal services meet and that the proper steps be taken to prevent rust or corrosion. Bolts and screws must be done up as tightly as possible and maximum contact obtained at conduit fittings.

The sizes of conductors to be used as earth continuity conductors are given in Tables D2 and D3 of the IEE Regulations, and if the e.c.c. is made up entirely of copper, a copper alloy or aluminium, it should have a maximum resistance of 1 ohm. If, however, it is made up partly of steel conduit or pipe, the maximum resistance permitted is $\frac{1}{2}$ ohm or 1 ohm, according to the test used.

The simple double-wound transformer

A transformer is a piece of apparatus without moving parts that transform alternating voltage and current supplied to one winding into different values of alternating voltage and current, usually in another winding at the same frequency.

Fig. 6.31 shows schematically a single-phase double-wound transformer consisting of two coils of insulated wire wound on a laminated core. The transformer core comprises the whole of the magnetic material forming the main magnetic circuit. In this type of construction the core is built up first and the coils of the two windings, which are usually circular, are arranged concentrically on the legs or limbs of the core. The remaining parts of the magnetic circuit are its yokes.

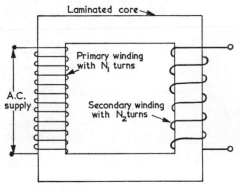

Fig. 6.31 Single-phase double-wound transformer.

The winding connected to the a.c. supply is known as the primary winding. It is on the input side of the transformer and electric power is supplied to it at the available alternating voltage.

The secondary winding is connected to the load. It is placed on the output side of the transformer and delivers electric power at another more convenient alternating voltage, at the same frequency.

Core and winding of larger transformers are immersed in mineral oil contained in a sheet-steel tank fitted with cooling fins, tubes or radiators.

The principle of the transformer

A transformer works on the principle of electromagnetic induction. Whenever an electric current flows in a coil of wire a magnetic field is set up which is concentrated along the axis of the coil. Whenever a coil of wire is linked by a changing magnetic field there will be a voltage set up in the coil. The simplest transformer consists of two windings of insulated wire on a single core of ferro-magnetic material.

When one of the windings (primary) is connected to an alternating current supply a current will flow in it. This causes a magnetic field to be set up by a changing current. This changing magnetic field is intensified by the core, and links the other winding (secondary) on the core. As a result a voltage is set up in the secondary winding which may be used to supply a load.

Transformer on open circuit

When an ideal transformer, that is, one with no losses, has its primary winding of N_1 turns connected to the supply and its secondary winding of N_1 turns on open circuit (see Fig. 6.32) a small sinusoidal alternating current is taken by the primary winding. This current,

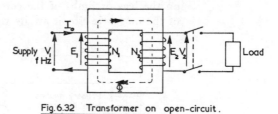

Fig. 6.32 Transformer on open-circuit.

whose r.m.s. value is I_0 amperes, produces $I_0 N_1$ ampere turns to set up a sinusoidal magnetic flux (see Fig. 6.33) with a maximum value of Φ_m webers and a frequency of f hertz in the magnetic core. The rate of

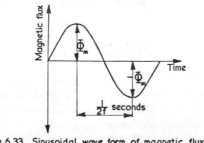

Fig. 6.33 Sinusoidal wave form of magnetic flux.

change of the alternating magnetic flux linking each turn of both windings is the same for both windings, and consequently the e.m.f.s induced per turn in both windings are the same. Let the induced e.m.f.s in the primary and secondary windings repectively have r.m.s.

values of E_1 and E_2 volts. These e.m.f.s are induced in opposition to the applied voltage of r.m.s. value V_1 volts. It can be proved that the values of these induced e.m.f.s are given by the following expressions:

$$E_1 = 4\cdot44\ \Phi_m f N_1 \text{ volts}$$
$$E_2 = 4\cdot44\ \Phi_m f N_2 \text{ volts}$$

The induced e.m.f. E_1 in the primary opposes the applied voltage of r.m.s. value V_1. Their difference is the voltage drop due to the primary no-load current I_0 flowing through the primary winding of the transformer. This voltage drop is ignored in the ideal transformer, and therefore the induced e.m.f. E_1 equals the applied voltage V_1.

∴   neglecting the primary voltage drop,

$$V_1 = E_1$$

∴ $V_1 = 4\cdot44\ \Phi_m f N_1 \text{ volts}$ **(6.28)**

Also, on no-load the r.m.s. value of the secondary terminal voltage V_2 equals the r.m.s. value of the secondary induced e.m.f. E_2.

∴ $V_2 = E_2$ on open circuit

∴ $V_2 = 4\cdot44\ \Phi_m f N_2 \text{ volts}$ **(6.29)**

From expressions 6.28 and 6.29

$$\Phi_m = \frac{V_1}{4\cdot44 f N_1} = \frac{V_2}{4\cdot44 f N_2} \text{ webers}$$

∴ $$\frac{V_1}{N_1} = \frac{V_2}{N_2} \text{ volts/turn}$$

This means that the volts per turn is the same value for both primary and secondary windings.

Note that the expression $\Phi_m = \dfrac{V_1}{4\cdot44\ f N_1}$ webers gives the maximum value of the alternating flux in the magnetic core. Since the applied voltage V_1 and the frequency f are the system voltage and frequency, they are usually constant, and the number of primary turns N_1 is also usually a fixed value. Therefore since the values of V_1, f and N_1 are constant then the value of Φ_m must also be constant.

Transformer on load

If the load circuit is closed, a secondary current of r.m.s. value I_2 will flow through the complete secondary circuit (see Fig. 6.34). This current produces $I_2 N_2$ ampere turns in the secondary windings; that is, an

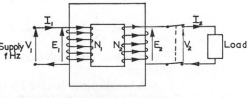

Fig.6.34 Transformer on load.

additional magneto-motive force (m.m.f.) which tends to set up more magnetic flux in the transformer core.

However, according to the expression $\Phi_m = \dfrac{V_1}{4 \cdot 44 \, f N_1}$ the maximum magnetic flux in the magnetic core must be a fixed value of Φ_m since V_1, f and N_1 are constant. Therefore a current I'_1 will be induced in the primary winding so that the ampere turns $I'_1 N_1$ neutralise the m.m.f. of $I_2 N_2$ ampere turns due to the load current I_2. The current I_1 in the primary winding will then be resultant current due to I'_1 and I_0, since the total current I_1 must also provide $I_0 N_1$ ampere-turns to produce the constant alternating flux of maximum value Φ_m in the core. The ampere-turns $I_0 N_1$ are small, and if neglected an approximation of $I_1 = I'_1$ can be made. There will also be a voltage drop in the secondary winding when the load current I_2 flows through it, but this voltage drop is neglected in the ideal transformer so that the induced secondary e.m.f. E_2 is taken as being equal to the secondary terminal voltage V_2 on load.

Therefore, in the ideal transformer, the magnetising ampere-turns $I_0 N_1$ and the primary and secondary winding voltage drops are neglected. Then,

$$I_1 N_1 = I_2 N_2$$

and
$$V_1 = E_1 = 4 \cdot 44 \, f \Phi_m N_1$$
$$V_2 = E_2 = 4 \cdot 44 \, f \Phi_m N_2$$

$$\therefore \quad \frac{V_1}{V_2} = \frac{E_1}{E_2} = \frac{N_1}{N_2} = \frac{I_2}{I_1}$$

$$\therefore \quad \frac{V_1}{V_2} = \frac{N_1}{N_2} = \frac{I_2}{I_1} = k \qquad (6.30)$$

Since $\dfrac{V_1}{V_2} = \dfrac{I_2}{I_1}$ then $V_1 I_1 = V_2 I_2$

$$\therefore \quad \frac{V_1 I_1}{1000} = \frac{V_2 I_2}{1000} \qquad (6.31)$$

Expressions 6.30 and 6.31 are the important relationships which are used in most transformer calculations at this stage, and they must therefore be memorised.

The ratio $\dfrac{V_1}{V_2}$ is known as a voltage ratio

ratio $\dfrac{N_1}{N_2}$ as a turns ratio, and

ratio $\dfrac{I_2}{I_1}$ as a current ratio. The constant k is sometimes called the transformation ratio of a transformer.

A transformer is called a step-down transformer if its secondary voltage V_2 is lower than its primary voltage

V_1; that is, if the transformation ratio is greater than unity.

Thus for a step-down transformer the high voltage winding is the primary winding and the low voltage winding the secondary winding. In symbols, for a step-down transformer

$$V_1 > V_2 \text{ i.e. } k > 1$$

A transformer is a step-up transformer if its secondary voltage V_2 is higher than its primary voltage V_1; that is, if the transformation ratio is less than unity.

Thus for a step-up transformer the high voltage winding is its secondary winding and the low-voltage winding its primary winding. In symbols for a step-up transformer

$$V_1 < V_2 \text{ i.e. } k < 1$$

Equation 6.31 gives the rating of the single-phase double wound transformer in kilovolt-amperes or since these relationships have been obtained for a single-phase transformer, the rating of each of a polyphase transformer (a transformer of more than one phase). Thus the primary and secondary quantities are the primary and secondary phase values respectively for a polyphase transformer.

The product $V_1 I_1$ is the number of volt-amperes that the primary winding has been designed to transform, and similarly $V_2 I_2$ is the number of volt-amperes that the secondary winding has been designed to deal with continuously without overheating. Thus

$$\frac{V_1 I_1}{1000} = \frac{V_2 I_2}{1000}$$

is the rating in kilovolt-amperes for which the transformer has been designed.

Exercises

1 What is **a** an electromotive force
  b an electrolyte
  c an electrode?

2 What is the difference between a primary cell and a secondary cell? Describe the action of a simple primary cell.

3 Explain the terms 'local action' and 'polarisation'. How are these defects overcome in a practical primary cell?

4 Describe the construction and action of either **a** the Daniell cell, or **b** the Leclanché cell, pointing out the advantages and disadvantages of the cell you select. Upon what factors does the internal resistance of the cell depend?   (W.J.E.C.)

5 Give the connection diagrams of a battery comprising six 2 V cells to provide **a** a 12 V supply
  b a 6 V supply
  c a 2 V supply

6 Describe, giving connection diagrams where necessary, any experiment you have carried out with cells connected **a** in series, **b** in parallel. (W.J.E.C.) Mention any conclusions you reached.

7 a Explain the difference between a primary cell and a secondary cell

   b When a lead-acid accumulator is being tested what are the indications that (i) it is in a fully charged condition, (ii) it is in a discharged condition? (W.J.E.C.)

8 What is electromagnetic induction? State Faraday's laws of electromagnetic induction.

9 Make a simple sketch of a single-loop generator and explain how a sine wave e.m.f. is induced. Sketch the sine wave and indicate the maximum and r.m.s. values on the sketch. (W.J.E.C.)

10 A three-phase 6·6 kV alternator has it star-connected winding earthed at the star point, and supplies a three-phase balanced load using a four-wire, three-phase system. Sketch the wiring diagram for the above system and indicate the line to line and line to neutral voltages on the diagram. (W.J.E.C.)

11 Explain with the aid of a sketch the action of a simple double-wound transformer. Give two examples of the use of such a transformer in electrical installation work.

12 Calculate the respective turns in each winding of a simple single-phase double-wound transformer that has a step-down ratio of 480 V to 240 V if the volts per turn are 1·6.

7
Use of conductors and insulators

Engineering materials
Most of the materials used in engineering fall into one of two important groups. Current will flow easily in the materials of one group and they are known as conductors. Current will not flow easily in materials of the second group and they are called insulators.

The dividing line between the two groups is not always clear because there are some materials which act as conductors under one set of conditions and as insulators in another. An example of this is paper which will act as an insulator when dry and as a conductor when wet.

Using conductors and insulators
Electrical engineering depends in equal measure upon conductors and insulators. Conductors are needed where electricity is generated, to carry it from the generator to where it is used and in the equipment which uses it. At all points in this chain insulators are also needed to ensure that current only flows where it is wanted and can be used without danger to life or property. The following table contains some of the materials frequently used in electrical engineering.

Conductors	Insulators
Copper	Rubber
Aluminium	Polyvinyl chloride (PVC)
Brass	Porcelain
Silver	Glass
Lead	Cotton
Carbon	Paper

These are of course only a few of the very large number of conductors and insulators that are available.

Although all conductors have in common the ability to carry an electric current they vary a great deal in other ways. Some are rigid, some are flexible, some may be compressed and not stretched. Solid conductors have different reactions to heat and different melting points and some conduction takes place in liquids and gases. Similarly whilst all insulators are alike in offering a high resistance to the flow of electricity they may be very different in other ways. The conductor and insulator that is used for a particular purpose is very often selected more for its other qualities than for its purely electrical characteristics.

Cables

Conductors are used to move electrical energy from one place to another. The best conductors from the point of conductivity and cost are copper and aluminium. In many cases the conductor is made up in a fairly flexible form with its own insulation and protection in which case it is known as a cable. Different types of cable are distinguished by the conductor material used, the insulating material and the protective covering if any.

(a) circular stranded

(b) stranded sector shaped

(c) solid sector shaped

(d) concentric

Fig7.1 Typical cable conductor shapes

A conductor and the insulation covering it make up a cable *core*. In some cables the conducting part is solid and in others it is stranded. For single core cables the conductor is usually circular but if there is more than one core the conductors may be shaped so that the overall cross-section of the cable is circular. In yet another cable arrangement one of the cores is placed outside the other. These are known as concentric conductors.

Cable sizes

The current which a cable can carry is determined by the size of its conductors. It is therefore important to understand how these sizes are stated. There are two possibilities: one is to refer to the nominal cross-sectional area of the conductor and the other is to quote the number of strands and their diameter. For example, the smallest type of cable used for fixed wiring could be called a 1·0 mm², meaning that its conductors have a cross-sectional area of approximately one square millimetre. It could alternatively be called a 1/1·13 (single-one-one-three) meaning that its conductors have one strand 1·13 mm in diameter. Similarly one of the larger cables could be called a 4·0 mm² or a 7/0·85 (seven-nought-eight-five).

Types of cable

The various types of cable are called by names which are simply groups of letters. These letters are in fact a short description of the various layers of insulation and protection which cover the conductor. Each letter is the initial letter of a layer. This will become clearer as particular cables are considered. The following are some of the cables in common use.

V.R.I. taped and braided

The initials in this case stand for 'vulcanised rubber insulated' and taped and braided refers to the additional layers (Fig. 7.2i). This type of cable was until recently

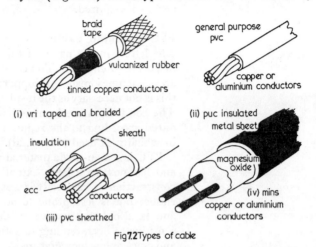

Fig 7.2 Types of cable

very widely used particularly in conduit installations. At the present time its use for new installations has fallen off considerably, but it is found so often in existing installations that it is worth considering.

It consists of a stranded tinned copper circular-sectioned conductor. This is covered with a layer of vulcanised rubber varying in thickness from 0·75 mm on the smaller sizes to 1·25 mm on the larger sizes. These thicknesses are for low-voltage cables, for medium and high voltages greater thicknesses are required. The vulcanised rubber is covered with a braided cotton or jute sleeve. This is coated with a moisture- and weather-proof compound.

The copper conductor strands are tinned to prevent the copper coming into contact with the vulcanised rubber which contains sulphur which would corrode the copper.

Pure rubber would not be suitable for cable insulation as it absorbs moisture and changes its state quickly. It must therefore be vulcanised by the addition of sulphur and by the application of carefully regulated heat. This makes it much more resistant to water and also tougher and less prone to deterioration.

Originally the purpose of the cotton tape was to keep the rubber in shape between application and vulcanising.

More up-to-date methods however make it possible to vulcanise the rubber as soon as it is formed over the conductors. The cotton tape therefore is only useful from the standpoint of any additional protection it gives the cable. Since this is slight it is normally omitted from the smaller sizes of cables.

Pigments are added to the rubber and the compounding to give the cable whichever colour—red, black and so on—is required.

PVC insulated cable

The letters in the name of this cable are a contraction of polyvinyl chloride. The conductors are stranded and are sometimes made of copper; it may be tinned but this is not necessary as the insulation contains no sulphur. The conductors are insulated with PVC which is coloured according to the requirements of the installation in which it is used (Fig. 7.2ii).

PVC is a man-made material which resembles rubber and has now to a large extent replaced rubber in the construction of cables. It is not as elastic as vulcanised rubber but is less prone to attack by acid and alkalis and is also less sensitive to oil and oxygen. PVC is sensitive to heat, tending to soften at high temperatures and crack at low temperatures.

Both PVC insulated cables and VRI taped and braided need additional protection such as conduit or trunking when they are installed.

PVC insulated and sheathed cable

For surface work PVC insulated cable with a toughened PVC sheath is frequently used. The sheath is generally grey in colour. It provides sufficient extra protection for cables used in homes, offices and similar situations where the risk of mechanical damage is not very great (Fig. 7.2iii).

This type of cable is available in a number of different forms. Sometimes the sheath only contains one core, sometimes a core and a non-insulated earth continuity conductor (e.c.c.) and sometimes it contains a number of cores with or without an e.c.c. Single-core cables have an overall circular cross-section whilst the multi-core ones are usually flat.

MIMS cable

This cable is very different from those already described although it is often used instead of PVC sheathed cables. The letters stand for mineral insulated metal sheathed and its construction is illustrated in Fig. 7.2iv.

The conductors are made of solid copper or aluminium

and the cable may contain one, two, three, four or seven of them. They are insulated from each other and the sheath by highly compressed magnesium oxide. This in the conditions in which it is used in the installed cable has a very high insulation resistance. The insulation and conductors are enclosed in a continuous copper or aluminium sheath which makes it a very robust cable. The magnesium-oxide insulation is hygroscopic, i.e. it is readily able to absorb moisture. Because of this a special method of sealing is needed for MIMS cable to prevent moisture entering and causing short circuits. The copper or aluminium sheaths have a good resistance to corrosion but where necessary they can be provided with a PVC oversheath.

This is an extremely useful type of cable, being suitable for many industrial situations, particularly where high temperatures are encountered, as well as for domestic, commercial, church and other installations.

Armoured cables

Cables used where there is a serious risk of mechanical damage are often protected by steel wires or tape. Such cables are said to be armoured and are widely used as buried supply cables and for general industrial work.

Of the armoured cable already installed the greater part is known as p.i.l.c.s.w.a. (Fig 7.3i). As usual the initials which form the name indicate the various layers. There are rather a lot of layers but each has its function and is therefore quite necessary.

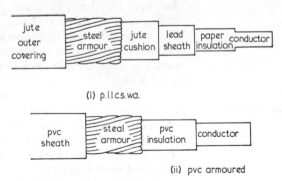

(i) p.i.l.c.s.wa.

(ii) pvc armoured

Fig 7.3 Armoured cables

The conductors are usually stranded untinned copper and may be circular or sector-shaped in cross-section. Each conductor is insulated with several layers of oiled paper tape and all the conductors are then formed into a group with an overall binding of oiled paper tape making the cross-section of the cable circular. Since the insulating properties of the paper tape are reduced by water the next layer takes the form of a continuous lead sheath. This prevents any moisture in the atmosphere reaching the cable cores and also prevents the oiled paper drying out. The lead sheath is too soft to be exposed to the risk of mechanical damage so it is served with jute and then protected with a spiral binding of steel wire or tape. The jute serving acts as a cushion and prevents the steel wire from cutting through the lead sheath. The steel armouring itself is then protected by a further serving of bitumen-compounded jute to prevent rust or corrosion. This is heavily dusted with chalk to reduce the tendency of turns to stick together when the cable is on its drum.

The tendency at the present time is to use less paper insulated cable and to use instead PVC insulated armoured cable such as that shown in Fig. 7.3ii. The particular cable shown has solid sector-shaped aluminium conductors insulated with general purpose PVC compound coloured according to the phases. The cores are grouped together by an extruded PVC sheath which is covered by a spiral steel-wire armouring which is in turn covered by a PVC outer sheath.

Cables of this type are obviously much simpler in construction than the paper insulated ones previously described and for this reason they are cheaper to manufacture and install. The major saving on the installation side is due to the absence of the lead sheath which makes plumbed joints unnecessary. They are also flexible and lighter than their p.i.l.c.s.w.a. equivalent (and less dirty) to handle. The insulation of PVC cables is more stable as far as moisture is concerned, but on the debit side its temperature sensitivity must be considered. Electrical protection of PVC armoured cables is more difficult than in the case of paper insulated cables.

Flexible cords

Sometimes a smaller type of insulated conductor or group of conductors than those that have been described is needed. For example to connect socket outlets to portable appliances or to connect a moving machine part to a fixed part. Since they are liable to frequent movement they must be more flexible than the cables already discussed. If such a cable has conductors with a cross-sectional area of less than 6 mm² it is called a flexible cord. To obtain the additional flexibility needed the conductors of a flexible cord have a larger number of strands of smaller diameter than is the case with ordinary cables. Cord sizes vary from 0·5 mm² (16/·20) to 4 mm² (56/·30), anything larger and having small cross-section strands being termed a flexible cable. The number of cores in a flexible cord will depend upon the use to which it is put.

Types of flexible cord

Flexible cords can be divided into three main classes. These are parallel twin, twin-twisted and circular cords or flexes, as they are often called. Each class can be further divided due to variations in their construction and the materials used for insulation and covering. The following are some of the types of cord most frequently encountered.

PVC parallel twin

This is a very small cord consisting of two cores insulated with PVC. Sometimes each core has a separate covering but in other types the covering is in one piece as shown in Fig. 7.4i. This type of cord is only suitable for the smallest loads such as electric shavers and similar appliances.

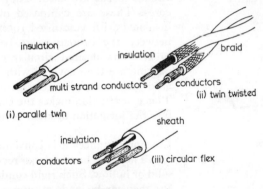

Fig 7.4 Flexible cords

Twin-twisted flex

This type of cord is used in a large number of cases as the connecting link between a ceiling rose and a lamp-holder. There are several varieties. The basic version (Fig. 7.4ii) consists of two stranded tinned copper conductors separately insulated with vulcanised rubber. Each core is then given an individual covering in the form of a cotton or artificial-silk sheath. The cores are then lightly twisted together—hence the name. The outer sheaths are frequently maroon in colour and the vulcanised rubber insulation will be red on one core and black on the other.

The development of compact lamps and the use of close-fitting shades has resulted in a general increase in the temperatures of light fittings. To compensate for this, flexes for lighting have been developed using butyl-rubber and glass fibre for insulation. These can sustain much higher operating temperatures and consequently have a much longer working life than those insulated with vulcanised rubber.

Circular flexible cords

These are cords in which the cores are suitably packed and sheathed so that the flex has an overall circular cross-sectional area. The most widely used of these are sheathed with tough rubber and PVC respectively and are called TRS circular flex and PVC circular flex (Fig. 7.4iii). The smaller sizes are used as lighting flexes in bathrooms and other places where additional protection is necessary. The larger sizes are used for hand lamps, electric drills and all sorts of portable tools and machinery.

Some types of portable appliances are moved much more than others; examples of these are electric irons, kettles and hair dryers. For such equipment *kink proof* circular cords are used. They may have two or three cores. These are composed of tinned copper strands insulated with vulcanised rubber or PVC. The spaces between the insulation are packed with cotton cords to give the desired circular cross-section. Cores and packing are then covered with a thin elastic sheath and an overall covering in the form of a tight cotton braid. This construction makes the cord rather stiff and therefore the formation of kinks is more difficult.

Bus-bars

In some instances it is convenient to use rigid conductors. They may be circular or rectangular in cross-section, solid or hollow. Such rigid conductors are called *bus-bars* and they can be made of copper or aluminium (Fig. 7.5).

The largest bus-bars are used extensively in sub-stations for interconnecting switchgear. As sub-stations

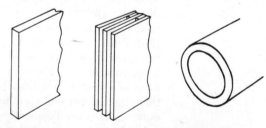

Fig. 7.5 rectangular solid   multi-bar   circular hollow

are inaccessible to unauthorised persons the bus-bars used in them are not generally insulated. In another form they are used to provide a flexible distribution system for machine shops and similar locations. Bus-bars when so used are enclosed in steel cases with points for connecting fused 'tap in' boxes at short intervals. This is known as bus-bar trunking. Vertical bus-bars are used to take supplies upwards in multi-storey buildings; these are known as rising main systems. On a smaller scale bus-bar chambers are used for ease of connection when making up distribution panels for commercial and industrial situations.

Ratings of cables, cords and bus-bars

Whenever a conductor carries current it warms up. If it gets too hot it is obvious that damage can result. The larger the current that flows the greater will be the rise in temperature that results. A conductor with a large cross-sectional area will carry more current without overheating than a conductor with a small cross-sectional area. It follows therefore that as the current to be carried increases so the size of conductor increases. The current which a conductor can carry under normal working conditions is called its *current rating*.

The IEE regulations contain tables which give recommended *current rating* for cables, cords and bus-bars in common use. It must be remembered that a cable does not have a *current rating* that is constant for all conditions. This is because the ability of a conductor to dispose of the heat generated in it by a current depends on the way in which it is installed. Accordingly the IEE tables make allowance for such factors as whether cables are run in conduits or on the surface, singly or in groups. The basic *current ratings* quoted in the tables are liable to be modified for certain variations from standard conditions. These variations are known as 'rating factors' and the basic 'current rating' is multiplied by them. Rating factors apply when the air temperature around the conductors varies from the normal (30°C). They may also be required by the use of 'coarse' or close excess-current protection, the grouping or arrangement of cables and the type of sheath used.

Terminations and connections

Whenever a conductor is connected to another conductor or to a piece of equipment a terminal or connector is used. There are very many different sorts and a few of the most common are illustrated in Fig. 7.6.

The back entry terminal shown is widely used for lighting switches and appliances. Various sizes of the rectangular screwed block terminal are used for a considerable range of switches and isolators. The split post arrangement is found in junction boxes where cables enter from a number of different directions. The shrouded connector is for joining smaller insulated cables. For laboratory work where connections are

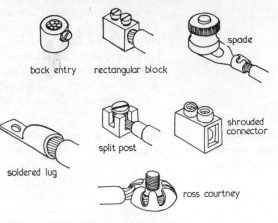

back entry   rectangular block   spade

shrouded connector

split post

soldered lug

ross courtney

Fig 7.6 Terminals and connectors

frequently changed insulated terminal post and spade connections are convenient. The spade terminal may have a soldered or crimped connection.

Multi-stranded cables in control panels and similar locations are frequently terminated with Ross Courtney washers which hold all the strands firmly. For larger cable terminations soldered or crimped lugs are needed and clamps of various kinds are used for bus-bar connections.

Conduit and sheathed wiring

All of the wiring used inside buildings requires protection in one form or another. Where the risk of mechanical damage is slight the protection may be in the form of a toughened PVC, or less frequently rubber, sheath. Cables for installations where the risk is greater can be sheathed in copper or aluminium. In industrial installations the protection may take the form of trunking or conduit. Conduit is the general name used to describe various forms of tubing. The most widely used type is black enamelled mild steel with threaded joints. Where corrosion is more likely galvanised steel is used. In other situations PVC conduit is used.

Summary of regulations

Cables: IEE Regulations Part 2 A11, A30, A33, A50, A66–69, B1–78, B85, B86, B102, B116, B124–139, C6, D3, D28, D29, F10, G15, G18–25, H1, J8, J12–16, K12, K15, K18, K20, K21, K26.

Cords: IEE Regulations Part 2 A10, B4, B5, B8, B10–18, B33, B36, B51, B52, B54–59, B73, B78, B128–132, C20, C27, D28, J8, K12, K26.

Bus-bars: IEE Regulations Part 2 B2, B21, B37, B56, B82, F1.

Termination and connections: IEE Regulations Part 2 B60–73, B78, F10, K26.

1 Select the best conductor from the following list:
a carbon, b silver, c steel, d brass

2 Which of the following conductors is used for overhead lines?
a bronze, b tin, c aluminium, d brass

3 Which of the following insulators is most affected by heat?
a glass, b paper, c cotton, d PVC

4 Which of these insulators is affected by direct sunlight?
a rubber, b porcelain, c PVC, d paper

5 Which of the following is used in mineral insulated cables?
a carbon dioxide, b magnesium oxide, c lead sulphate, d polyvinyl chloride

6 Which of the following statements are correct:
a VRI taped and braided cable is used less frequently now than it was in the past   True/False
b Mineral insulated cables should not be used in warm places   True/False
c The cores of flexible cables have more strands than standard cables of a similar cross-sectional area   True/False
d A flexible cord has conductors with a cross-sectional area not exceeding 6 mm^2 True/False
e The current rating of a cable depends only upon its construction and its cross-sectional area   True/False

7 Describe with the aid of diagrams the construction of three different types of cable in common use.

8 Give a brief description of the ways in which sizes of cables are stated.

9 What are the relative merits and disadvantages of paper insulated and PVC insulated armoured cables.

10 Draw cross-sectional diagrams of three different types of flexible cord and state where they would be used.

11 Examine a section of copper or aluminium bus-bar and draw sketches to show how lengths are joined together, supported and how connections are made to it.

12 What factors affect the current ratings of cables?

13 Draw a sketch showing how a PVC sheathed cable is terminated at the outgoing side of a single-phase distribution board.

14 Draw neat sectioned sketches showing how cables are jointed using
a insulated connectors and b a joint box with split post terminals.

8
Current-using equipment

Heating appliances

Heat can be transmitted by conduction, convection and radiation. Conduction is the term used to describe what happens when heat travels through a solid object. If an object is heated the air or liquid directly above it is heated, lightens and rises. This is then replaced by cold air or liquid which is in turn heated and rises. As a result a continuous current of air or liquid known as a convection current is set up and eventually all the air having access to the object becomes heated. If metal is heated until it glows, energy waves are transmitted outwards warming anything that lies in their path. This is known as radiation. Most space-heating equipment makes use of radiation or convection. Some use only one form whilst some makes use of both.

Radiant heaters

These are probably the most widely used electrical heating units and are available in many forms. Radiated heat warms the bodies on which it falls without greatly affecting the air through which it travels. Its effects, however, are immediately felt, unlike those of convection, and this fact tends to make radiant heaters very popular. In their most common form they make use of firebar elements or rod-type elements.

In the firebar-type heater the energy source is a high-resistance wire wound in a spiral about 6 mm in diameter and set in grooves in a block of fire clay (Fig. 8.1). Passing current through the wire causes it to become red hot and in consequence it radiates heat. The temperature of the fireclay also rises until after a short while it starts to radiate. From then on the whole surface of the bar radiates heat. The air above the fireclay warms up and so additional heat is given off by convection. About 60% of the total heat energy given off is radiated and 40% is by convection.

In rod and reflector heaters the element is wound on a fireclay rod about 20 mm in diameter. This is mounted at the focal point of a highly polished reflector (Fig. 8.2). The element is heated to incandescence by passing current through it and it thus radiates heat. Some of this heat travels straight out from the element but the greater part of it is caught by the reflector and concentrated into a strong beam which may be directed as required. Because the rod is small and the reflector remains fairly cool, only about 20% to 30% of the heat given off by the rod and reflector heater is in the form of convected heat.

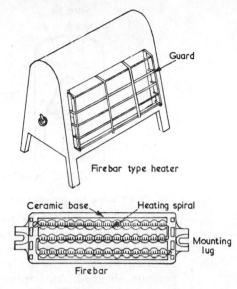

Firebar type heater

Ceramic base — Heating spiral

Mounting lug

Firebar

Fig. 8.1

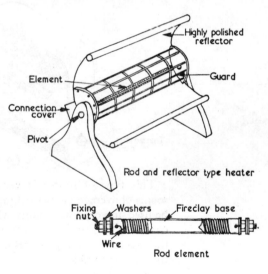

Highly polished reflector

Element

Guard

Connection cover

Pivot

Rod and reflector type heater

Fixing nut — Washers — Fireclay base

Wire

Rod element

Fig. 8.2

Convection heaters

In its simplest form the convection heater consists of a tube open at both ends and containing a heating element (Fig. 8.3). When the element carries current it warms up and heats the air in the tube which rises and leaves the tube. The pressure in the tube is reduced as a result of the warm air leaving and this allows cold air to enter at the lower end which in turn is heated and so the process continues. The hot air on leaving rises

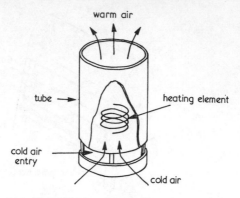

Fig 8.3 Simple convection heater

to the highest point in the room and there displaces cooler air which is moved to a lower level. The resulting circulation of air means that eventually all parts of the room will be affected by the heater and a uniform temperature obtained. An improved variety of the basic heater is shown in Fig. 8.4.

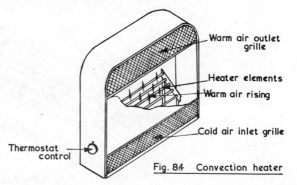

Fig. 84   Convection heater

This type of heater is very efficient when used in rooms of average ceiling height. It should not, however, be used in very lofty rooms where the air is frequently changed. The more sophisticated types may contain a thermostat operating according to the temperature of the air at the intake.

Fan heaters

In a fan heater, which is a variation of the basic convection heater, the flow of air over the heating element is boosted by a fan powered by an electric motor. This allows the flow of heated air to be directed as required and also greatly increases the quantity of air that can be heated in a given time. A simple domestic heater is illustrated in Fig. 8.5. Fans of a more robust pattern are frequently used for heating workshops. A common variation of the industrial type uses an electrically driven fan to supply air to a hot-water heater.

140

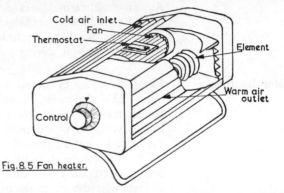

Fig. 8.5 Fan heater.

Storage heaters

These are generally used as part of central-heating systems in older buildings, as they can be installed without any alteration to the structure. There are a number of different types of storage heater available but they are generally similar in construction. A fairly typical one is illustrated in Fig. 8.6. The elements in such a heater are enclosed in substantial blocks of dense heat-storing material such as furnace brick. The core is enclosed in a layer of thermal insulation and protected by a sheet-steel container.

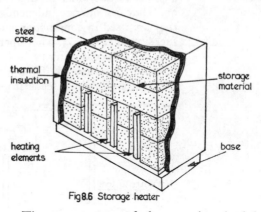

Fig 8.6 Storage heater

The temperature of the core is raised by passing current through the heating elements. This process raises the temperature of the core about 400°C. Once the core has warmed up, the heat it stores is given off slowly through the thermal insulation and case. The surface temperature of the case does not normally exceed 100°C.

From this it will be seen that the operation of the storage heater has two easily defined phases: one phase during which it absorbs heat energy and a second phase during which it gives off its stored heat. The maximum advantage from the unusual characteristics of these heaters is obtained when 'off-peak' tariffs are available.

An 'off-peak' tariff allows electrical energy to be purchased at reduced rates when the demand is low. The heater uses the electricity to heat it up when it is cheap and gives off heat during other times.

Water heaters

Large quantities of heated water are used every day in industrial and domestic installations. A wide choice of appliances are available for water heating and the choice of any particular type depends on the purpose for which the heated water is required.

There are three standard methods by which water is heated electrically. The first is by means of a heating element placed beneath the container in which the water is held. The second is by means of an insulated and moisture-proof element completely surrounded by the water it heats. This is known as the immersion method. The third method is by passing currents through the water to be heated between electrodes at different potentials. The second method is very widely used and so we will concentrate on appliances making use of it.

Pressure-type water heaters

These are used in most homes to obtain the bulk supplies of hot water needed for baths, washing and so on. They can operate independently, but in many cases they are used in conjunction with the boiler of a solid-fuel system (Fig. 8.7). They operate as follows: The feed tank delivers cold water to the bottom of the heater tank. The tank is completely filled with water which also fills the vent pipe which leads from the top of the heater

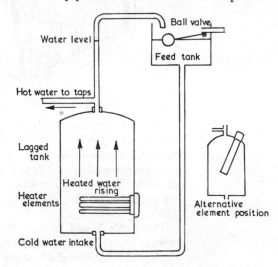

Fig. 8.7 Pressure-type water heater

tank back to the feed tank. When the heater elements are switched on the water heated by them rises to the top of the heater tank back to the feed pipe. The hot-water taps fed by the heater are connected to a pipe leading off from the junction of the heater tank and the vent pipe. When one of the hot-water taps is opened heated water is drawn off and the level of water in the system falls. This causes the feed-tank ball valve to open, allowing water in from the mains to restore the water-level. The temperature of the water in the tank is controlled by means of a thermostat associated with the heating elements.

For maximum efficiency the heating elements should be placed near and parallel to the bottom of the tank. In practice, however, it is often convenient to fit the element into the top of the tank. This is called a pressure-type water heater because the water is delivered at a pressure which depends upon the 'head' of the water in the feed tank. The degree of efficiency achieved with this type of heater depends to a considerable extent on how well the tank and hot-water pipes are lagged to prevent unnecessary heat loss.

**Free-outlet
water heater**

Where it is not convenient to install a feed tank and all the associated pipe-work a 'non-pressure' or 'free-outlet' heater may be used. The arrangement is shown in Fig. 8.8 and can be connected directly to a cold-water pipe.

The action of this type of heater is as follows: The water in the tank is heated by the elements which are controlled by an external switch and an internal thermostat. Once the water surrounding the elements is heated, it rises to the top of the tank. When the control tap at

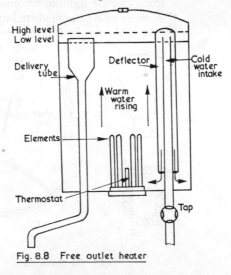

Fig. 8.8 Free outlet heater

the intake is opened cold water is delivered to the bottom of the tank. This raises the general level of water. The heated water at the top of the tank then overflows and runs off through the outlet which is always open and gives this type of heater its name.

Both this and the pressure-type heater belong to a larger group known as storage heaters. They are so called because they keep a supply of heated water ready for use.

Instantaneous heaters

Where the amount of hot water required is not great, or the demand infrequent, 'instantaneous heaters' are used. These may work on the electrode principle and do not store any hot water, but heat as required. The amount of electrical energy used by this type of heater in a given time tends to be fairly large, but none of it is wasted in keeping water hot.

Electric lamps

Electric lamps fall into one of two main categories. In one type light energy is released as a result of heating a solid until it glows by passing current through it and in the other it is released by passing current through a suitable gas. The first type are known as filament lamps and the second as discharge lamps.

Filament lamps

Work on these began over a century ago, with experimental lamps having platinum filaments. These early lamps had no practical value and progress was slow until the carbon filament was developed. Because it could sustain a very high temperature the carbon-filament lamp was quite efficient when working in a glass bulb from which the air had been removed.

Once electric lighting became a commercial proposition, every effort was made to improve on the early crude lamps. This resulted in the gradual evolution of the

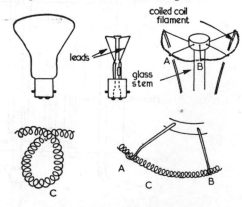

Fig8.9 The filament lamp

modern filament which operates in a glass bulb filled with inert gas. The purpose of the gas is to prevent the filament from boiling away at the high temperature needed to give a good light output. The gas tends to cause a reduction of efficiency through cooling and this is offset by means of the coiled-coil arrangement shown in the diagram. The bayonet cap (BC) arrangement shown is generally quite satisfactory for small lamps (up to 150 W). Some lamps, particularly the larger sizes, have screwed caps. These are known as Edison Screw (ES) and Goliath Edison Screw (GES) respectively.

Discharge lamps

When a metal is heated a cloud of electrons will form about it. An electrode at a higher potential placed near the heated metal will attract some of these electrons. The greater the potential difference, the greater will be the flow of electrons. The heated metal and other electrode can be enclosed in a glass tube. The heated metal is called the cathode and the electrode at high potential is called the anode. If the tube is filled with a gas such as mercury vapour collisions will take place between the electrons moving from cathode to anode and some of the gas molecules. These collisions release extra electrons which also move towards the anode with a resultant increase in current. The energy released as a result of the collisions appears in the form of light. Electric lamps which make use of this principle are called discharge lamps.

Fluorescent lamps

Probably the most common form of discharge lamp is the mercury-vapour fluorescent lamp. This type of lamp requires a special circuit which is shown in Fig. 8.10. Its action is as follows.

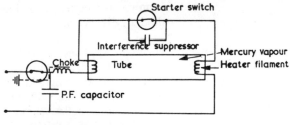

Fig. 8.10. Fluorescent lamp circuit

When the a.c. supply to the lamp is switched on the contacts of the starter are closed and current flows through the filaments at each end of the tube. They heat up and an electron cloud forms. After a short time the starter contacts open automatically. This breaks the flow of electrons which causes a high voltage to be set up in the choke. This voltage is applied across the ends of the tube, attracting electrons from the filament at the lower potential to that at the higher potential.

In moving from one end of the tube to the other, the electrons collide with many mercury-vapour molecules. The collisions release energy in the form of ultra-violet radiation which strikes a coating of fluorescent powder on the wall of the tube making it glow brightly. The collisions also release extra electrons to join in the flow and cause further collisions. A gas molecule which has lost an electron has a positive charge and is attracted to the electrode at the lower potential. The positive gas molecules strike the electrode hard enough to keep it hot and allow extra electrons to join the cloud around it.

Once the breakdown of gas molecules starts, the current tends to increase rapidly. Damage to the tube would result from this if a volt drop across the choke did not limit the voltage available to pass current through the tube. The capacitor across the mains is needed to improve the *power factor* of the circuit. The capacitor across the starter switch is to prevent the opening and closing of the starter causing radio interference.

Other discharge lamps

For outside lighting particularly for streets and roadways two other types of lamp are especially suitable. These are the *sodium vapour* and the *high-pressure mercury vapour* lamp. They combine a high light output with freedom from shadows and glare. Unfortunately both of them give light which has a bad effect on colours. In industrial installations ever-increasing use is being made of the *quartz iodine* type of lamp which has an even greater light output per watt. For decorative and advertising work the cold-cathode discharge lamp is used. They are better known as 'neon' lamps because neon gas is frequently used as the medium in which the discharge is set up.

Summary of regulations

Heating Appliances: IEE Regulations Part 2 A1, A58–62, B36, B133–139, F1, F2.

Lamps and Lampholders: IEE Regulations Part 2 A56, C22–26, C28, C39, D9, D16, F3, K25.

Lighting Installations and Fittings: IEE Regulations Part 2 A25, A58–62, B70, B132, C16, C27, D1, D6, D9, D16, F1, F3, J17, J18.

Discharge Lighting: IEE Regulations Part 2 G1–6.

Exercises

1 Which of the following statements is correct:
  a Convection is what takes place when heat travels through a solid   True/False
  b Radiated heat does not have much effect on the air through which it travels   True/False

c The heating effect of a radiant heater is not felt as quickly as that of a convection heater   True/False

d Off-peak tariffs make cheap electrical energy available at all times of the day   True/False

e A rod-type element forms the most important part of most convection heaters   True/False

2 Select the correct answer to each of the following questions from the alternatives provided.

a In what type of room should convection heaters not be used?

(i) small rooms, (ii) rooms with low ceilings, (iii) rooms with very high ceilings, (iv) bedrooms

b Which type of water heater requires a separate cold-water tank?

(i) Free outlet, (ii) pressure type, (iii) instantaneous, (iv) electrode boiler

c Why should the tank and pipework of a water-heating system be lagged?

(i) to reduce vibration, (ii) to limit noise, (iii) to reduce heat losses, (iv) to reduce the risk of electric shock

d What material is used to manufacture the filament of a modern electric-light bulb?

(i) carbon, (ii) tungsten, (iii) glass, (iv) mild steel.

e Why is a capacitor connected across the starter switch of a fluorescent lamp?

(i) to prevent it overheating, (ii) to limit radio interference, (iii) to assist the flow of current in the filaments, (iv) power factor correction

3 State the type of electrical heating which you would select for a living room in use for only two hours a day. Give reasons to justify your choice.

4 What are the relative advantages and disadvantages of radiant and convection heaters?

5 Briefly describe the construction and operation of a storage type of electrical space heater.

6 What factors should be taken into consideration when designing a water-heating system for a small family dwelling?

7 Describe with the aid of diagrams what happens in a free outlet water heater following the turning of the tap to obtain a supply of hot water.

8 How can the heat losses which occur in a domestic water-heating system be kept down to a reasonable level?

9 Draw a neat sketch to illustrate the construction of a large filament lamp. Label the components and state the purpose of each.

10 Describe the construction and operation of a mercury-vapour fluorescent lamp.

9
Electrical diagrams

In this chapter we shall briefly discuss electric circuits and wiring diagrams, the conventional symbols for common components and the colour coding used for electrical components and wiring. It is assumed that draughtsmanship, use of instruments, execution of drawings, interpretation of sections and first angle projections have been taught elsewhere in the syllabus of the subject 'Communication'.

Electric circuits and wiring diagrams

As mentioned earlier in Chapter 4, page 47, all electric circuits have three essential parts, namely:

    1 the source

    2 the conductors between the source and load

and 3 the load

The source is that part of the circuit that produces electrical energy at the expense of some other form of energy, e.g. chemical, mechanical, or thermal.

The conductor system is the part of the circuit between the source and the load. It consists of insulated metal conductors in the form of cables and switches or switchgear.

The load is the part of the circuit that utilises the electrical energy by reconverting it to either chemical, mechanical or thermal energy.

An electrical-circuit diagram is a diagram that shows the various constituents or components and their interconnection in the operation of the circuit by means of graphical symbols. The aim of the circuit diagram is to show the operation of the circuit as clearly as possible. The conductors or wires of the circuit are shown as straight lines drawn parallel to the edges of the paper and the various pieces of electrical equipment represented wherever possible by standard graphical symbols. For the sake of clarity, a circuit diagram does not necessarily represent the actual physical relationship in space between the various parts of a circuit and their connections.

A wiring diagram is a diagram that shows the wiring between components or items of equipment and may show their layout. It also makes easier the checking of internal and/or external connections. The connections made between components or items of equipment may be given on a wiring diagram in tabular form.

A wiring layout is a drawing that shows the physical layout of the wiring. One should note that any type of symbol may be used on any type of diagram if it improves the clarity or understanding of the diagram.

The method of representation used in any of the above diagrams may be according to the number of conductors, devices or elements represented by one symbol. It may be *multi-line* representation, where each item of apparatus or each element is represented by one symbol and each conductor by an individual line. In *single-line* representation two or more conductors are represented by one single line and similar items of equipment by one single symbol. See Fig. 9.1 for some of these stan-

Fig.9.1

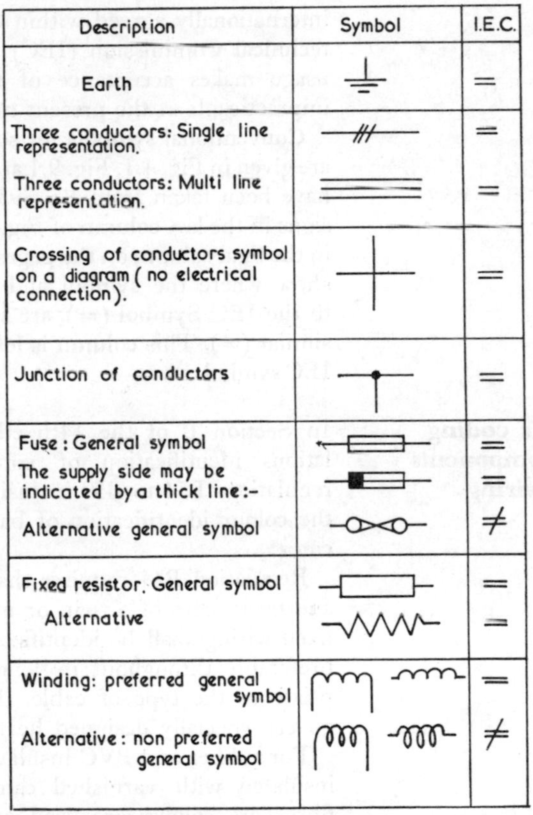

Description	Symbol	I.E.C.
Earth		=
Three conductors: Single line representation.		=
Three conductors: Multi line representation.		=
Crossing of conductors symbol on a diagram (no electrical connection).		=
Junction of conductors		=
Fuse : General symbol		=
The supply side may be indicated by a thick line :-		=
Alternative general symbol		≠
Fixed resistor. General symbol		=
Alternative		=
Winding: preferred general symbol		=
Alternative : non preferred general symbol		≠

dard symbols. Alternatively representation may be according to the relative placing of the symbols on the diagram for different parts of an installation which corresponds wholly or partly to the relative physical location of the objects for example, an architectural diagram.

149

Conventional symbols for common components

The British Standard 3939: 1966 'Graphical symbols for electrical power, telecommunications and electronics diagrams' gives the standard symbols that should be used in electric circuits and wiring diagrams. It is a combination and revision of BS 108: 1951 'Graphical symbols for general electrical purposes' and BS 530: 1948 'Graphical symbols for telecommunication and supplements' 1–7.

BS 3939 will supersede both BS 108 and BS 530. Appendix 8 to the 14th edition of the IEE Regulations for the Electrical Equipment of Buildings lists some symbols for use in wiring diagrams and location symbols for installations from BS 108. It is intended to revise Appendix 8 in due course to take account of BS 3939. The symbols used in BS 3939 are identical with those internationally agreed within the International Electrotechnical Commission (IEC) except where established usage makes acceptance of the international symbol impracticable at the present time.

Conventional symbols for some common components are given in Fig. 4.1, Fig. 9.1 and Fig. 9.2. These symbols have been taken from BS 3939 and BS 108. Note the signs in the last column of Fig. 9.1. These are signs used in the final column on the pages of symbols of BS 3939 to show where the symbol and description are identical to the IEC Symbol ($=$), are similar to ($\approx$) or are dissimilar ($\neq$). This column is left blank when no relevant IEC symbol exists.

Colour coding for components and wiring

In Section B of the 14th edition of the IEE Regulations identification of conductors is covered by regulations B54 to B59 inclusive with Table B4 giving the colour identification of bare conductors and cable cores.

Regulation B44 requires that every single-core cable and every core of a twin or multicore cable for use in fixed wiring shall be identifiable at its termination and preferably throughout its length by a method appropriate to the type of cable. This regulation does not govern specially designed heating cables.

For rubber and PVC insulated cables and for rubber insulated with varnished cambric and heat-resisting fibre core colours are used or appropriately coloured discs or sleeves at terminations. These recommended colours are listed in Table B4.

For armoured PVC insulated cables as well as the above methods an alternative method uses numbered cores in accordance with BS 3346; 1961 'Armoured PVC insulated cables' provided that the numbers 1, 2 and 3 indicate live conductors, the number 0 the neutral

Fig.9.2

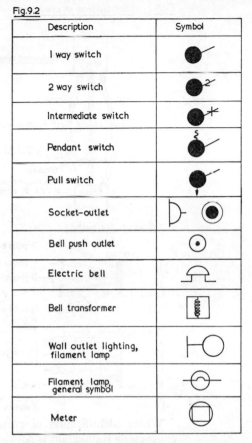

Description	Symbol
1 way switch	
2 way switch	
Intermediate switch	
Pendant switch	
Pull switch	
Socket–outlet	
Bell push outlet	
Electric bell	
Bell transformer	
Wall outlet lighting, filament lamp	
Filament lamp, general symbol	
Meter	

conductor and the number 4 is given to the special-purpose fifth core if present in the cable.

For paper and varnished-cambric insulated cables and for varnished p.t.p. fabric insulated cables the cores are numbered in accordance with the relevant British standards provided that numbers 1, 2, and 3 denote live conductors, the number 0 the neutral conductor and the number 4 indicates the fifth (special-purpose) core if present in these cables.

For mineral insulated cables the method used is the application of appropriately coloured sleeves or discs in accordance with Table B4 to the ends of the cable.

These various methods are illustrated in Fig. 9.3.

Regulation B55 recommends that bare conductors be made identifiable where necessary by means of appropriately coloured sleeves or discs in accordance with Table B4 or by painting with those colours. This regulation is not applicable to bare earthing leads or to bare earth continuity conductors or to the earth-continuity conductors in complete cables.

Fig.9.3   Identification of conductors

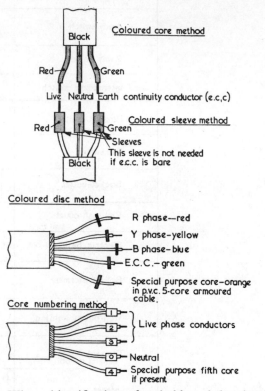

Where identification of switchboard bus-bars in a consumer's installation is not covered by Table B4 then by Regulation B56 their means of identification shall comply with BS 158: 1961 entitled 'Marking and arrangement of switchgear bus-bars, main connections and small wiring'.

The cores of every flexible cable, including flexible cord, by Regulation B57 shall be coloured throughout in accordance with the requirement of Table B4. The exception to this regulation is PVC insulated parallel twin non-sheathed flexible cords. This type of cable to BS 2004 is governed by Regulation B58 which stipulates that the core with the longitudinal rib shall be used for the live conductor or the d.c. positive or outer conductor (see Fig. 9.4).

Fig.9.4

Also government regulations have stipulated recently that the colours of the cores in three-core flexible cables or cords attached to domestic electrical appliances such as electric irons, refrigerators, etc., sold in future shall be:

green and yellow   for earth continuity conductor
brown   for live conductor
blue   for neutral conductor

This new colour code has been adopted by Great Britain and eighteen other European countries to avoid the danger of incorrectly connecting plugs to any domestic electrical appliance bought in any of the nineteen countries.

After the 1st November 1970 for manufacturers and wholesalers, and the 1st April 1971 for retailers, government regulations make it illegal for them to sell any domestic electrical appliance fitted with a three-core flexible cable or cord coloured differently to the new colour coding. Existing three-core flexible cables or cords with the old colours, namely, green for the earth core, red for the live core, and black for the neutral core, may still be used. A useful publication that gives information on the new colour code and how to correctly connect the new flexible cables to plug-tops is a card entitled 'The New Wiring Colours'. It is issued by the Home Office and Scottish Home and Health Department and may be obtained free of charge from most Electricity Board showrooms and Her Majesty's Stationery Offices.

Regulation B59 requires that in any scheme for distinguishing electric conduits from the pipes of other services in buildings, the conduits shall be coloured light orange in accordance with BS 1710: 1960 entitled 'Identification of pipelines'.

Exercises

1 The symbols in List 1 must be matched with the appropriate components in List 2. Record your answer by writing the appropriate number in the match panel provided.

LIST 1

A —|⊢—
B —o—o—
C —⋁⋁⋁⋁—
D ⊥

LIST 2

1 Fuse
2 Primary or secondary cells
3 Variable resistor
4 Transformer
5 Earth

Match Panel

A	B	C	D

2 Identify the following BS graphical symbols.

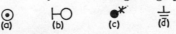
(a)   (b)   (c)   (d)

3 Make a neat sketch showing a 13 A fused plug connected to a three-core flexible cord, indicating the correct colour of each core. The sketch should show the plug top removed.   (W.J.E.C.)

4 Draw a circuit diagram of an electric bell circuit comprising a trembler bell, continuous action relay with electrical reset and push. The power supply to the bell circuit is from a double-wound fused transformer.

5 Give the BS graphical symbols for:
  a a ceiling outlet for discharge lamp
  b a socket outlet
  c a pendant switch
  d a bell-push contact
  e a single-phase double-wound transformer.

Part B

10

Lighting and heating circuits

Equipment at the consumer's intake point

The supply authority for any district will terminate its supply cable or lines at some suitable point on the consumer's premises. The form of this termination will vary according to the size of the consumer's installation. The place where the supply terminates is generally called *the intake*. Some of the equipment at the intake is provided by the supply authority and some must be provided by the consumer.

Supply authority's equipment

If a paper insulated armoured cable is used the first item of equipment at the intake is a *sealing box*. This keeps the moisture in the atmosphere from reaching the cable insulation and provides a protected enclosure in which the cable conductors can be connected to the 'tails' which replace them. Where a PVC insulated service cable is used or overhead lines a special junction box replaces the sealing box.

Once an installation is connected to the supply it is possible that any fault in the installation may affect the supply. To reduce the likelihood of this happening the area boards enforce the IEE Regulations as far as they are able. In spite of this some element of risk still exists and in order to guard against it a *service fuse and link* is installed. It is placed as near as possible to the intake. A typical service fuse and link or 'cut out' as they are often called is shown in Fig. 10.1.

The 'cut out' consists of a fuse of the correct rating for the particular consumer's installed load connected

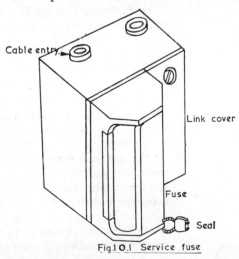

Fig.10.1 Service fuse

in the 'line' conductor and a solid bar in the neutral conductor. If an excess current large enough to overload the service cable flows without being cleared by the consumer's protective devices the service fuse blows. Damage to the service cable is thus prevented and the supply to other consumers is not affected. The bolted solid bar or link makes it possible to isolate completely the supply from the installation without disconnecting any wiring. This makes it possible to work on other equipment at the intake without any risk.

The other important item of equipment at the intake which belongs to the supply authority is the *energy meter*. This records the amount of energy consumed by an installation.

The energy consumed is a variable depending upon supply voltage, current flowing and length of time for which the current flows. The standard energy meter contains a rotating disc which is driven by the magnetic fields of two electromagnets. The strength of the field of one is determined by the voltage of the supply and that of the other by the current flowing. The rate at which the disc turns therefore depends on the power in the circuit at any given instant. The number of turns made will depend upon the time for which the power is maintained. The disc is mounted on a spindle which drives a train of gears and these record the resulting combination of power and time. The resulting energy is recorded in kilowatt hours.

The items previously considered are installed by the supply authority, that is, the local area board. They remain their property and the consumer must not interfere with them in any way.

Consumer's equipment

For compliance with the IEE Regulations the consumer must make certain provisions at the intake point. These may be summarised as follows. The consumer must provide for the disconnection of the installation in the event of a dangerous earth leakage. He must also provide for its disconnection automatically if excess current flows and lastly there must be provision for complete isolation of the consumer's part of the installation from the supply in order to carry out repairs or alterations.

The simplest way of meeting these requirements is by the use of a main switch fuse. The fuse provides protection against excess current and also against earth leakage whilst the switch mechanism provides the necessary isolation. For this arrangement to be satisfactory the current flowing in the event of an earth fault must be large enough to blow the fuse. The impedance of the earth path must therefore be low.

An alternative method is to use a combined over-current and earth-leakage circuit breaker. This method has the advantage that it can function in conditions where the impedance of the earth path would be too great to allow a fuse to blow when an earth fault develops. It is however rather more expensive than is the switch fuse.

If it is felt to be an advantage separate units can be used to meet each requirement. In this case the isolator must be next to the consumer's terminals, that is immediately following the supply authority's meter. A simple arrangement at the intake is illustrated in Fig. 10.2.

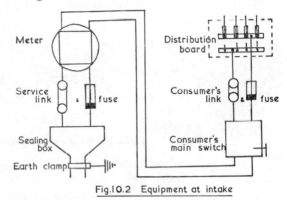

Fig.10.2 Equipment at intake

After the main control gear the consumer will place distribution boards feeding the final sub-circuits of the installation. A final sub-circuit is that part of an installation between the last control or fuse and the apparatus to which electricity is being fed. A distribution board is a unit containing a set of fuses or circuit breakers.

Main excess current protection is not necessary if the following requirements are met:
a The cables between the service fuse and the consumer's distribution board or boards have at least the same rating as the service fuse and
b Either the isolator is in the same enclosure as the distribution fuses or very close to it.

This exemption makes it possible to use consumer control units which considerably simplify and tidy the intake point. The ratings and capacities of all the switchgear fuses and circuit breakers used at the intake point must be adequate for the currents they are likely to carry or interrupt under normal or fault conditions.

Domestic and office distribution

The conditions that exist in homes and offices are quite similar. Installations in them are not subject to extremes of temperature nor any great risk of mechanical damage nor any of the special hazards encountered in industrial situations. The basic difference is one of scale, the cables and accessories used being for the most part the same.

Domestic installations

The electrical installation in a home has to provide for lighting, heating, cooking, water heating and portable appliances. In most cases this is done by making use of a single-phase a.c. supply. Distribution in domestic installations is based on the consumer's control unit. This is a combined main switch and distribution board and the outlets from it may be protected by fuses or miniature circuit breakers (Fig. 10.3). The most common

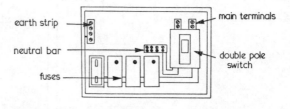

Fig10.3 Interior of consumers control unit

arrangements have four, six or eight outlets or ways but units with larger numbers of ways are available if needed. The main reason for the popularity of the consumer unit apart from simplifying arrangements at the intake is that the rating of each way can be varied according to the requirements of the installation. This gives it a degree of flexibility which the ordinary distribution board does not have. A typical domestic unit might contain two ways rated at 5 A for lighting, one at 15 A for a water heater, one at 30 A for a cooker, one at 30 A for a ring circuit and a spare way to allow for future requirements.

Office installations

For smaller offices these tend to be similar to domestic installations but with the accent more heavily on lighting and heating. Larger offices, however, require a three-phase supply to deal with the increased load, although most of the individual items of equipment will be designed for single-phase operation.

A suitable arrangement for such an installation is indicated in Fig. 10.4. The supply after the usual intake controls and metering is taken into a three-phase and neutral fuse-switch. This has high rupturing capacity fuses to give overall short-circuit protection. The outgoing side of the switch is connected to a bus-bar chamber mounted directly above it. Tappings are taken off the bus-bars; in this case they serve three distribution boards, one for lighting, one for heating and one for socket outlets and odd small appliances. The loads are spread as far as possible over the three phases to give the best possible balance. Arrangements such as this

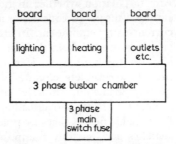

Fig 10.4 Simple office control arrangement

can be designed to control medium and large installations but when the installation is spread over a number of floors a rising main system is often used.

A rising main system consists of a set of vertical bus-bars running from the bottom to the top of the building they serve. A typical section of rising main is illustrated in Fig. 10.5. The supply for each floor in

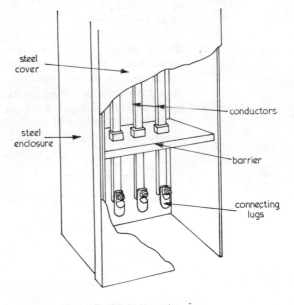

Fig 10.5 Rising main

turn is tapped off from these bus-bars. Multi-storey office blocks frequently make use of open-plan arrangements to allow for maximum flexibility in laying out the working area. In order to make supplies available wherever they are needed with such arrangements, a grid of under-floor ducting may be used. This ensures that cables can be run to within a few feet of any point on the floor without difficulty.

Practical circuitry

Any installation whether in home, office or factory is simply a means of connecting a supply of electricity to current-using equipment. In order to do this a complete circuit is needed from the source through the load and back to the source. It is not, however, sufficient to provide a complete circuit. The circuit must be adequately insulated, suitably controlled and protected against the effects of overloading short circuits and earth faults. This means that any practical circuit will contain switches and possibly thermostats as well as fuses, or overload and earth-leakage circuit breakers.

Switches, thermostats and circuit breakers may be designed to break a single conductor or more than one. If they break only one they are described as single-pole, if they break two or more than two they are known as double-pole and multi-pole respectively.

Polarity

All public supplies have one point earthed as has been mentioned in Chapter 6. In a two-wire installation one of the conductors originates at the earth point and is known as the neutral conductor. There will be no significant difference in potential between the neutral conductor and any earthed metalwork associated with the installation. The other conductor in the two-wire installation is called the line. Single-pole switches, circuit breakers and thermostats must not be connected in the neutral conductor nor must fuses, as it is dangerous to do so. The danger of connecting these devices in the neutral can best be appreciated by considering Fig. 10.6.

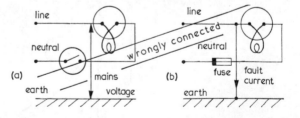

Fig 10.6 Dangers of incorrect polarity

In the part (a) of this diagram it can be seen that when the switch is in the neutral conductor it can be switched 'off' but still leave the mains voltage present between the lighting point and any earthed metal. This could obviously be very dangerous for anyone working on the lighting point. In part (b) of the diagram a fuse is shown in the neutral conductor. If an earth fault occurs under these circumstances the fault current will not flow through the fuse and it will not 'blow'. In addition to this if the fuse is removed the circuit will cease to function but there will still be a supply at the lighting point. If it is desirable to make provision for breaking the neutral, it can be made by providing a solid bolted link which cannot be removed whilst the line conductor is complete. Devices connected in the right conductor are said to have correct *polarity* and those in the wrong conductor are said to have *incorrect polarity*.

Lighting circuits

There are two main classes of lighting circuits. Those which make use of metal filament lamps and those using discharge lamps. At this stage only filament lamp circuits will be considered.

If a filament lamp is to light it is necessary that a current should flow through it. The current must be large enough to heat the filament and cause it to give a good light output. On the other hand the current must not be so great that it causes the filament to overheat and burn out. In order to meet these requirements it follows that the lamp used must be suitably rated for the supply to which it is connected. In all the circuits that follow it will be assumed that lamps and supplies are suitably matched.

In its simplest form a lighting circuit could consist of a lamp connected directly across the supply. This is obviously not practical as the supply would have to be cut off to turn off the lamp. Any practical circuit must have some means of control and this usually takes the form of a single-pole switch; it will also need to be protected against excess current and earth leakage.

A single lamp circuit is illustrated in Fig. 10.7. Note that the *line* side of the supply is connected to the switch

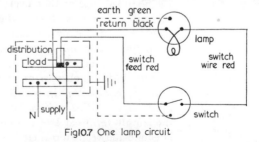

Fig10.7 One lamp circuit

and the *neutral* side of the supply to the lamp. The cables used should be coloured as shown. An earth connection is required at every lighting point and switch.

Such a circuit with only one lamp is unusual, most circuits contain more than one lamp and more than one switch. Additional lamps are connected in parallel with the supply. Where the circuits contain more than one switch the supply to each must be independent.

The following circuits are the ones most commonly used for connecting groups of lamps and switches.

Yard-lighting system

This system as the name implies is used for lighting schemes in factory yards and other places where runs of pole-mounted lights are used. The circuit is shown in Fig. 10.8. The basic requirement is that line, neutral and

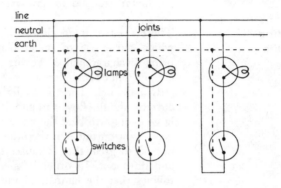

Fig.10.8 Yard lighting system

earth wires are run out to the furthest lighting point. They are tapped at each light as shown in the diagram with a single-pole switch in the lead from the line conductor. This system can be employed both with overhead lines and with underground cables.

Looping-in system

A disadvantage of the yard-lighting system is that it involves making joints at every tapping point. This is obviously unsuitable for indoor work as the IEE Regulations require that every joint should be readily accessible. For this reason when cables are to be protected by conduit or trunking it is usual to adopt the 'looping-in' method. In this system each switch and lighting point acts as a junction box. This is illustrated in Fig. 10.9 which also gives the names used for the different conductors in the circuit. The circuit is built up as follows. The switch feed connects the line side of the supply terminal 1 of switch A. A second conductor is connected from terminal 1 of switch A to terminal 1 of switch B and from there to terminal 1 of switch C. This is known as looping and provides an uninterrupted path from the line side of the supply to each switch. The neutral conductor from the supply is connected into terminal 1 of lamp A and looped into lamps B and C.

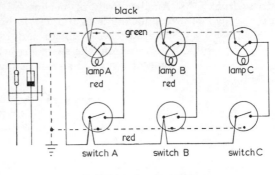

Fig10.9 Looping in system

This provides an uninterrupted return path from each lamp. The circuit for each lamp is completed by connecting switch wires from terminal 2 of each switch to terminal 2 on the corresponding lamps. An earth point is provided at each lamp and switch.

Three-plate ceiling-rose system

A disadvantage of the system just described is that in some cases it will be necessary to run three wires or if the shortest paths are taken only one wire. This is not desirable especially for surface wiring. To avoid this *three-plate ceiling roses* are used. A three-plate ceiling rose is one having an extra main-current-carrying terminal. This enables a lighting system to be wired throughout using a two-core and earth-sheathed cable (Fig. 10.10).

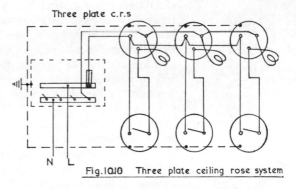

Fig.10.10   Three plate ceiling rose system

One cable, as shown in the diagram, runs through each lighting point and another runs from each lighting point to the switch controlling it. The switch feed is looped through the spare terminal at each ceiling rose and from there it is taken to the appropriate switch. The switch wire connects the outgoing side of the switch back to the lamp. The return to the supply is by way of the neutral conductor which is looped at the remaining ceiling rose terminal.

Loop-in switch system

Another way of dealing with the problem of odd conductors is to use *loop-in switches*. These are single-pole switches with an additional terminal completely separate from the switching action. A circuit using this type of switch is shown in Fig. 10.11. The underlying

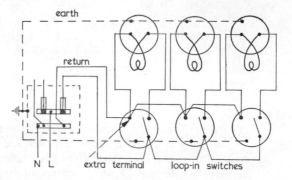

Fig 10.11 Use of loop-in switches

principle is the same as that of the three-plate ceiling-rose system, but in this case the main circuit cable runs through each switch position and auxiliary cables run to the lighting points from the switches. The spare terminal is used as a looping point for the neutral conductor.

Two-way switching

It is often useful to switch a light on at one point and off at another. For example someone using a corridor should be able to switch on and off at either end. For this type of control special switches with three terminals are needed. The switching mechanism connects a common terminal to one of the remaining terminals in one position and to the other in the second position. The way in which they are connected into a lighting circuit is shown in Fig. 10.12. Such switches are called *two-way* switches.

In Fig. 10.12a it can be seen that a closed circuit exists. Through this current can flow from the fuse in the distribution board through the switch feed and into switch 1 through the common terminal C. It leaves by terminal 1, flows through the upper strapper terminal 1 and the common terminal of switch 2, through the switch wire, lamp and neutral back to the supply. As a result the lamp lights up. In Fig. 10.12b the circuit is broken at switch 1 by switching over to link terminals C and 2 which puts out the light. The light may be put back on either by returning switch 1 to the position shown in a or by changing switch 2 so that its terminals C and 2 are connected as shown in Fig. 10.12c. The remaining 'off' position is illustrated in Fig. 10.12d.

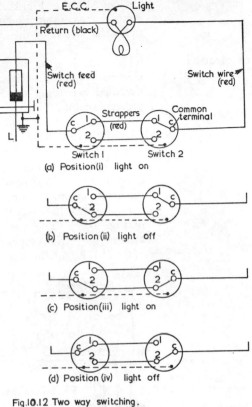

(a) Position(i)   light on

(b) Position (ii)   light off

(c) Position(iii)   light on

(d) Position (iv)   light off

Fig.10.12 Two way switching.

Extra lamps in circuits

In the circuits considered up to this point only one lamp has been controlled by each switch. In practice it is common for a switch to control several lamps. If this is the case modification of the circuits is necessary to provide a neutral and a switch wire at every extra lamp. This can be done in a number of ways as illustrated in Fig. 10.13. In Fig. 10.13a both switch wire and neutral

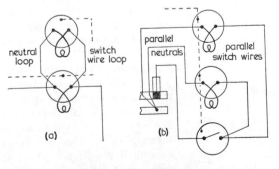

(a)   (b)

Fig 10.13 Connecting extra lamps

are looped and in Fig. 10.13 b they are paralleled. Either method or a combination of both may be used, depending upon the layout of the circuit involved.

Power circuits

This term is used to describe most applications of electricity apart from lighting. Such circuits are usually classified according to the rating of the fuse which protects them.

Circuits rated at 15 A or less

Where the fuse protecting a circuit is not rated at more than 15 A a number of units may be connected to it. The exact number will depend upon the current demand of each. For example three 5 A socket outlets could be fed by a single 15 A way in a distribution board as shown in Fig. 10.14. The current demands which must be assumed for various types of outlet are given in Table A3 of the IEE Regulations. .

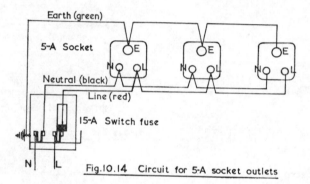

Fig.10.14   Circuit for 5-A socket outlets

Circuits rated at more than 15 A

With a few exceptions only one piece of equipment can be connected to a final sub-circuit rated at more than 15 A. In such a circuit a switch must be provided to break the line conductors or in the case of equipment with exposed heating elements, all the conductors. The switch can be part of the equipment as long as the equipment can be stripped for routine maintenance without exposing any live conductors.

Exceptions to the one-item rule

The one-item rule is relaxed in domestic installations to allow a circuit rated at not more than 30 A to supply more than one cooking appliance in one room. The kettle point on a cooker control unit does not count as a separate item. The one-item rule is also relaxed in controlled conditions where fused plug-tops are used with socket outlets.

The fused plug-top

If a 15 A socket outlet is installed it must be assumed that it will carry 15 A and its protective fuse rated accordingly. If therefore ten 15 A socket outlets are needed then ten sub-circuits must be installed. This would obviously require an excessive amount of wiring and control gear. Accordingly special regulations have been made to allow the connection of a fairly large number of socket outlets to single fuses. These regulations take into consideration the fact that in many installations only a small proportion of the installed outlets will be in use at any one time. To take advantage of these regulations special outlets are required which can only accept a fused plug-top. The fuse in the plug-top provides any protection needed by the appliance to which it is connected whilst the sub-circuit fuse guards against overloading of the circuit as a whole. For domestic installations the fused plug-tops used must be of the flat-pin 13 A type (Fig. 10.15).

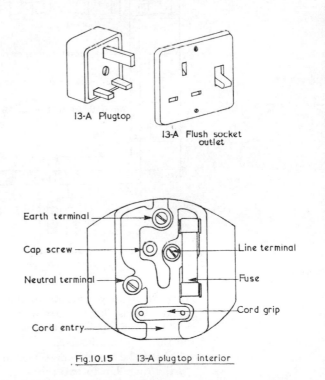

13-A Plugtop

13-A Flush socket outlet

Earth terminal

Cap screw

Neutral terminal

Cord entry

Line terminal

Fuse

Cord grip

Fig.10.15   13-A plugtop interior

There are two types of circuits making use of the flat-pin socket outlets and fused plug-tops. They are known as *ring circuits* and *radial circuits* respectively.

Ring circuits

In a ring circuit the line, neutral and earth conductors must be connected in the form of a loop, each returning to the terminal from which it started. This is illustrated in Fig. 10.16. Socket outlets are connected to the ring in such a manner that the loops are unbroken. With this arrangement each outlet is fed along two paths. The number of socket outlets that may be connected to such a ring circuit in a domestic installation is unlimited. One

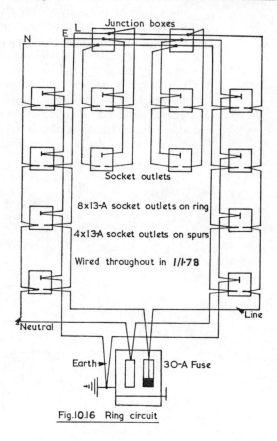

Fig.10.16 Ring circuit

ring circuit may serve up to 100 m² of floor area. For greater areas the number of rings must be increased and the total number of outlets spread fairly evenly over all the ring circuits used.

Not all the socket outlets need to be connected directly to the ring. Up to two thirds of the total number can be connected to spurs off the main ring. A spur is a run of cable tapped off the ring at a junction box or one of the socket outlets on the ring. A spur can carry two socket outlets or one fixed appliance not rated at more than 13 A and having its own switch and fuse. The total number of spurs must not exceed the number of socket outlets on the main ring.

If the cables with which the ring is wired are insulated with rubber or PVC they must have a minimum size of 1/1·78 mm (2·5 mm²). If it is wired with mineral insulated metal-sheathed cable the minimum conductor size must be 1·5 mm². The whole ring circuit is fed either from a circuit breaker, switch fuse, or one way of a distribution board or consumer's control unit. The maximum setting of the circuit breaker or rating of fuse used must be 30 A.

Radial circuits using 13 A socket outlets

In a radial circuit line, neutral and earth conductors are taken out from the distribution board and not brought back. This type of circuit will often be wired in two-core and e.c.c.-sheathed cable. Two socket outlets may be connected to a radial circuit wired in 1/1·78 mm PVC or rubber insulated cable or 1·5 mm² mineral insulated cable protected by a fuse or circuit breaker rated at 20 A. The same circuit could serve up to six socket outlets if they were all in the same room. The room, however, would need to have a floor area of less than 30 m², it must not be a kitchen and none of the outlets must feed a fixed water heater.

If from three to six socket outlets are required without any special conditions being observed, they may be connected to a circuit breaker or a fuse rated at 30 A. The wiring in this case must be either 7/0·85 PVC or rubber insulated cable or 2·5 mm² mineral insulated cable.

Summary of regulations

Equipment at intake: IEE Regulations Part 2 A1–7.
Polarity: IEE Regulations Part 2 A8, E2.
Final sub-circuits: IEE Regulations Part 2 A3, A23–55, E10.

Exercises

1 Complete the following sentences using one of the alternatives given:
  a Faults in an installation are prevented from affecting the supply by
      (i) the consumer's control unit, (ii) the cable sealing box, (iii) the service fuse
  b One of the things that must be provided by the consumer at the intake point is
      (i) a list of equipment connected to the supply, (ii) a means of isolating the installation from the supply, (iii) a distribution board for socket outlets
  c The requirements for equipment to be provided by

the consumer at the intake point are often met by installing

(i) a consumer's control unit, (ii) an energy meter, (iii) a single-pole isolator

d The most suitable way of supplying a multi-storey building is by means of

(i) PVC armoured cable, (ii) overhead lines, (iii) a rising main

e The neutral of an installation connected to an earthed public supply may contain

(i) a fuse, (ii) a bolted link, (iii) a single-pole switch

2 Choose the correct answers to the following questions from the alternatives provided:

a When wiring an installation what can be eliminated by using a 'looping-in' system?

(i) short circuits, (ii) tee-joints, (iii) single-core cables

b What is the name given to the conductor linking a switch and the lighting point it controls?

(i) control wire, (ii) switch loop, (iii) switch wire

c Which conductor is looped at the spare terminal in a three-plate ceiling rose?

(i) the switch feed, (ii) the neutral, (iii) the earth

d Which conductor is looped at the spare terminal in a loop-in switch?

(i) the switch feed, (ii) the neutral, (iii) the earth

e What type of sheathed cable is most suitable for use with three-plate ceiling roses?

(i) single-core, (ii) three-core, (iii) twin and e.c.c.

3 Describe the purpose of any two items of equipment that can be found at the point of a consumer's installation.

4 Why is a consumer's control unit frequently used at the intake point of an installation?

5 Describe an electrical distribution system suitable for use in a multi-storey office building.

6 Draw a neat diagram of a circuit consisting of four lamps each controlled by a single-pole switch and wired throughout using twin- and earth-sheathed cable.

7 Why should fuses not be installed in the earthed conductor of an installation?

8 Draw a neat diagram showing how to connect three pairs of lamps each pair controlled by a single-pole switch. Use the looping-in system and indicate cable colours and sizes.

9 Draw a diagram showing how a group of three lamps can be connected so that they can be switched on or off from either end of a corridor.

11
Wiring systems

Use of sheathed wiring

Much of the wiring used in offices, shops and homes is not exposed to any great risk of mechanical damage. It need not, therefore, be provided with armour or run in tubes or other enclosures. For such places sheathed wiring is used. It will most often consist of cables having PVC insulation and sheaths, but less frequently they may be insulated in general purpose rubber compound and have tough rubber sheaths. The construction of these cables is described in Chapter 7.

Sheathed wiring accessories

A wide range of accessories are available for use with sheathed wiring. Most of them are of the all-insulated type, a number of which are illustrated in Fig. 11.1. Some are designed for use in full view while others are intended to be at least partly conccaled.

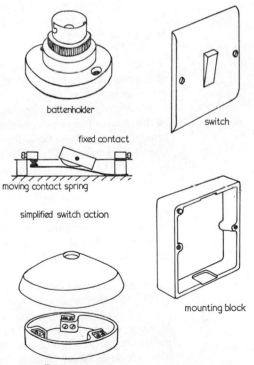

battenholder

switch

fixed contact

moving contact spring

simplified switch action

mounting block

ceiling rose

Fig11.1 All insulated accessories

Single-pole switches

The switch shown in the figure is of the type very frequently used. It is designed for flush mounting, in which case it is fixed to a metal box concealed in the plaster. If it is to be surface mounted a moulded box similar to that shown will be necessary. As most lighting circuits operate on a.c. supplies, a very simple switching mechanism can be used. This is called a micro-gap switch and is also illustrated.

Lampholders

The lamp in a lighting circuit is fitted into a lampholder which may be one of a number of different types depending on the requirements of the installation and the kind of lamp used. With sheathed systems bayonet cap lamps are most frequently used in conjunction with either a batten holder or a ceiling rose and a cord-grip lampholder.

The batten lampholder has the advantage of being a self-contained unit which is directly connected to the circuit wiring. In its basic form it consists of a base, barrel and lock ring with an additional shade ring if it is needed. It may be straight or angled, depending on whether it is to be mounted on a ceiling or wall. In order to protect the cables where their sheath is removed it must in most cases be mounted on a pattress.

The domestic rose (CR) is made of moulded bakelite. The main cables of the lighting circuit are terminated at the ceiling rose and a flexible cord is used to connect it to a lampholder. The terminals of the CR are designed to accommodate both cables and cords. The CR shown in the illustration is a three-plate one, that is it has three separate connection points for cable and flex. The earth terminal required is usually provided in the pattress on which the ceiling rose is mounted.

The cord-grip lampholder is so designed that the weight of the fitting and lamp is supported by the cord and not the lampholder terminals. Both cord-grip and batten lampholders may be fitted with bakelite skirts to prevent accidental contact with live conductors when lamps are being changed.

Joint or junction boxes

With a sheathed wiring system it is necessary to provide protection wherever any joints are made. This is because jointing involves breaking the sheath. The type of protection generally used is the all-insulated 'joint' or junction box. Some of these are plain and others have fixed terminal posts (Fig. 11.2, p. 175). With the plain variety insulated connectors are used. The sheath of the

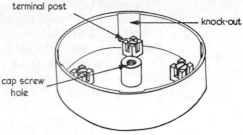

Fig11.2 Junction box

cable must be taken into the box, but not so far that it interferes with the joints or connections contained by the box.

Other equipment for sheathed systems

Apart from the small accessories already described there are a number of other things often used with sheathed cables. Probably the most common of these is the all-insulated consumer unit (Fig. 11.3). Also widely used is the all-insulated switch-fuse (Fig. 11.4). Less common is the all-insulated distribution board.

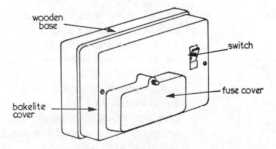

Fig11.3 All insulated consumers control unit

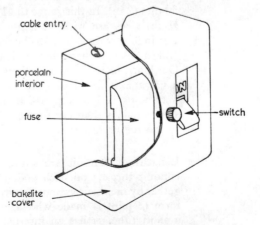

Fig11.4 All insulated switch fuse

Sheathed cables are secured by means of buckle-clips or moulded PVC clips fixed in position by brassed iron pins (Fig. 11.5). Where extra protection is needed as for example when the cables are run beneath plaster the usual procedure is to enclose sheathed cable in plastic oval conduit or cover it with galvanised steel or plastic troughing.

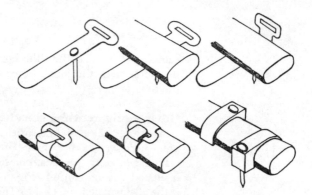

Fig11.5 PVC and buckle clipping

Conduit

In industrial situations such as factories or workshops one of the most popular methods of installation is to run unsheathed cables in screwed conduits. This system affords good mechanical protection. It also provides a degree of flexibility since once the conduit is erected cables may be removed or drawn in without any alteration to the line of the run. There are a number of different types of conduit. The ones most widely used are made of mild steel but conduits made of aluminium and PVC are also popular.

Conduits vary in diameter from 16 mm ($\frac{5}{8}$ in) to 50 mm (2 in). In this range the 19 mm ($\frac{3}{4}$ in) and 25 mm (1 in) sizes are without doubt the most widely used. For indoor work rolled-steel conduit with a welder joint and a black enamelled finish is extensively used. Outdoors, or in damp situations, the same conduit with a galvanised finish or solid drawn galvanised conduit may be used. It is supplied in straight lengths of about 4 m which is then cut to the required length and bent to the required shape.

Erecting conduits

Lengths of conduit are supplied with ready cut BS conduit threads on each end. Conduits are joined together by means of various screwed fittings. The simplest form of joint is made with a coupler which is simply a short tube. It has an internal thread into which the threaded ends of the conduits to be joined are screwed.

This type of joint and some other forms of straight-through joint are illustrated in Fig. 11.6.

Naturally runs do not always allow the use of whole lengths of conduit. In such cases the required lengths must be cut and threaded on site. To do this the conduit

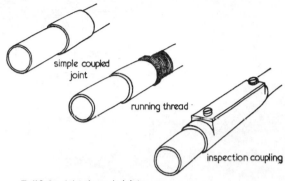

simple coupled joint

running thread

inspection coupling

Fig 11.6 Straight through joints

should be held in a suitable pipe vice. The greater part of the conduit's length should protrude from the back of the vice and the cutting point placed as near the jaws of the vice as possible. The piece to be cut off must be supported and the cut made right through. A fine hacksaw blade (12·5 teeth/cm) is used for this purpose. The thread is made with a set of stocks and dies as illustrated in Fig. 11.7. They should be in good condition,

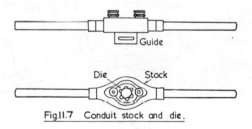

Guide

Die   Stock

Fig.11.7   Conduit stock and die.

clean and lubricated with a suitable cutting paste. The thread cut should be just long enough to occupy the whole of the receiving thread of the fitting into which it shall be screwed. After cutting, the end of the conduit will usually need to be filed square and then it must be reamed out to remove any rough edges which might damage cables as they are drawn in.

Not all of the joints are of the straight-through type. Accordingly it is often necessary to use one of the many conduit fittings that are available. Some of these are illustrated in Fig. 11.8. It is also frequently necessary to bend conduit. For this purpose a portable pipe-bending machine or occasionally a wooden bending block is used. The inner radius of a bend should not be less than 2·5 times the diameter of the conduit used.

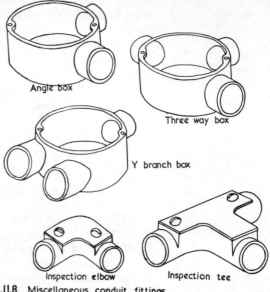

Fig.11.8 Miscellaneous conduit fittings.

Requirements for conduit installation

Conduit must be securely fixed and where necessary given additional protection against mechanical damage. The fixing devices most commonly used are shown in Fig. 11.9. The complete conduit installation must be erected before any cables are drawn in. Enough suitable

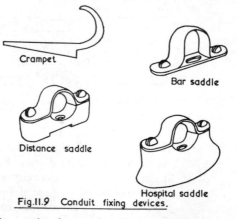

Fig.11.9 Conduit fixing devices.

inspection boxes should be provided to allow the cables to be drawn in easily and they must be accessible throughout the life of the installation. Solid tees and bends can only be used in two ways. First, they can be used next to inspection boxes, lighting fittings and similar outlets. Second, one solid bend can be placed in a run: under certain conditions that it is not more than 0·46 m (18 in) from the end of a run not more than 10 m (30 ft) in length and that the other bends in the run do not add up to more than one right angle. No run should contain more than two right-angle bends between draw-in points unless the number of cables they contain is reduced to facilitate drawing-in.

The number of cables that may be drawn into a conduit system is governed by Tables B5 and B6 of the IEE Regulations. If cables are not covered by these tables a space factor of 40% must be applied. This means that not more than 40% of the internal cross-section of the conduit can be taken up by cables. Slightly increased numbers of cables are allowable in short straight conduit runs such as those used for switch drops.

Avoiding damage to cables in conduit

The cables used in conduit systems are generally unsheathed and are therefore liable to damage. The most obvious source of danger is from rough edges of metal in the conduits. This is avoided by fitting smooth bushes wherever cables enter or leave conduits. The need to ream all cut ends has already been mentioned. Twisting and kinking of cables can also cause trouble. This can be avoided by exercising care and patience when drawing in. Damage can also be caused by dirt, small stones and other material such as cement in the conduits. It is necessary therefore to seal all outlets if there is any time lag between erecting conduits and drawing in wires.

Drainage and ventilation
There is always a risk of condensation in a conduit system. It is best prevented by adequate ventilation. This is provided by leaving sufficient openings at high and low points in the system to allow an uninterrupted flow of air. Any openings should be carefully sited so that no dust or falling moisture can enter. Any unnecessary entries should of course be plugged. Long runs of conduit should be angled slightly downwards with a drainage opening at the lowest points. Moisture can be kept out from joints and also rust can be prevented by painting any exposed thread at such places and thus seal the joint.

Identification
In factories and other places where conduits are likely to be confused with the pipes of other services a colour code is used. To comply with the appropriate Standard BS 1710 electrical conduits are painted orange either throughout their length or at regular intervals.

In large installations a stage may be reached where the cables to be installed are too numerous for the satisfactory use of conduits. At this point trunking is frequently introduced as a protective enclosure for non-sheathed cables.

The most widely used forms of trunking are square or rectangular in cross-sectional area with a detachable cover. They may be made from aluminium or mild-steel sheet or sometimes PVC.

The usual size range is from 35×35 mm to 100×150 mm, but larger sizes are available if needed.

A large range of tees, bends and other fittings are available to simplify changes of direction and inter-sections in Fig. 11.10. Because trunking cannot be set

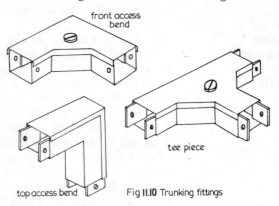

front access bend

tee piece

top access bend   Fig 11.10 Trunking fittings

there is a need for non-right-angle bends as well as the right-angle ones. This type of trunking is generally mounted in horizontal or vertical runs on walls. There are other types of trunking which are used for particular purposes.

Skirting trunking

Industrialised building methods make it difficult to get beneath the surface to conceal wiring. Skirting trunking is useful in such cases as it provides ample space and protection for wiring but is quite unobtrusive. It is available in steel or PVC. As the name implies it is run at the base of the walls of a room in the position usually occupied by the skirting board.

Floor trunking

In open plan offices, shops and factories it is necessary to be able to get a supply easily to any part of the working area. For this purpose floor trunking is very well suited. It is generally made of zinc-coated sheet steel and may only have one compartment or be sub-divided. Apart from the size the main difference between types of floor trunking is the form of covering used. In some cases they are covered with trays which can be filled with blocks or other material to match the overall floor finish.

Ducting: The term ducting applies to enclosures similar in construction and purpose to trunking but without a removable cover.

General notes

Insulated cables may be run in trunking without any further protection. If, however, there is any risk of mechanical or other damage the trunking should be given any necessary protection. Runs of trunking should be erected completely before any cables are installed in them. The number of cables installed should be such that a space-factor of 45% is not exceeded.

To avoid any strain on cables in vertical runs of trunking they should be supported at intervals of not more than five metres. Another risk arising in vertical ducts and trunking is of heat rising as a result of convection. The effects of this are limited by installing heat barriers in vertical runs at intervals of three metres and where the trunking passes through a floor. Similar barriers are required where horizontal runs pass through walls.

Mineral-insulated cables

These cables have been described in Chapter 7 as being made up of solid conductors insulated with compressed magnesium oxide and enclosed in a continuous copper or aluminium sheath. The insulation readily absorbs moisture and therefore special care is needed in sealing the ends of cables before they are connected to the supply.

Sealing mineral insulated cables

The parts which are used to seal a mineral insulated cable are illustrated in Fig. 11.11. They comprise pot, seal and tails. Before sealing is commenced a gland is fitted over the cable, the purpose of which is mentioned later.

brass pot   fibre disc   pvc tails

Fig11.11 Parts for sealing MI cable

In brief the sealing process is as follows. Firstly enough of the sheath is removed to disclose an adequate length of conductor for connection purposes. The insulation is removed and the conductors are cleaned.

The insulation resistance between conductors and sheath is then checked to ensure that there is no moisture present. The brass pot is then screwed on and filled with sealing compound. After that the insulating tails and seal are slipped over the conductors and crimped on to the pot to complete the seal. The insulation resistance between tails and to the sheath is then checked again and an identification sleeve is fitted to the tail which is to be used for the 'line' conductor.

The use of mineral-insulated cables

Mineral insulated cable may be used for surface work, concealed installations or it may be buried in the ground. It is most often seen in surface installations used in conjunction with conduit boxes and metal-clad fittings. When so used the cable is fitted to the conduit boxes or fittings by means of the gland mentioned in the paragraphs on sealing. There are two types of gland in common use. The first is made of solid brass, is cylindrical in shape and has a hole drilled through it. The hole is just large enough at one end to pass the sheath of the cable. The other end is bored out to a greater extent so that the gland can fit over their sealing pot. This type of gland has an external thread. The sheath is held in position by a pair of grub screws. The other type of gland consists of three parts of a body that screws into a conduit entry, a compression ring and a cap that screws on to the body and squeezes the ring down to bite on to the sheath of the cable and hold it firmly in place. A variation of the three-part gland with a longer conduit thread is used for flame-proof installations. The standard types of gland are illustrated in Fig. 11.12.

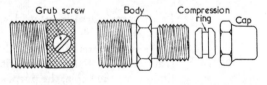

Fig.11.12 Mineral insulated cable glands.

Mineral insulated cable is most frequently used on the surface and when so run it is fixed with single or multiple clips. Some examples of these are shown in Fig. 11.13. If large numbers of cables are to be run the problem of fixing is simplified by the use of cable trays. Spacings for cable supports are given in Table B2 of the IEE regulations. The smallest bend must have a radius at least six times the diameter of the cable in which it is made.

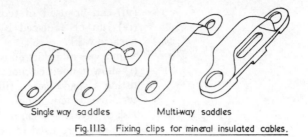

Single way saddles　　　　Multi-way saddles

Fig. 11.13　Fixing clips for mineral insulated cables.

Where these cables are run beneath the surface, the copper clips are replaced by PVC or other temporary fixings to hold the cable until the surface finish has set. If the material in which the cable is run is corrosive the cable is buried directly in made up ground.

All metallic conduit, trunking and the sheaths of mineral insulated cable must be earthed and generally they are used at least in part as the earth continuity conductor of the installation. It is important that all joints should be clean and solid to provide good electrical continuity. If non-metal boxes or fittings are used anywhere in such systems a suitable conductor must be used to link the metal parts. The total resistance of the e.c.c. must not exceed twice that of the largest conductor it protects.

Bonding to other services

Metal work of an installation should be kept clear of the metalwork, pipes etc. of other services. If this cannot be done bonding is necessary, that is, the metal of other services and of the installation should be connected together; this will reduce any danger resulting from earth leakage.

Summary of

Trunking: IEE Regulations Part 2 B35, B40, B42, B44, B49, B50, B53, B94, B111–118, B125, D3, D29.

Mineral-insulated cables: IEE Regulations Part 2 B1, B41–44, B53, B67, B85, D3, D29.

Conduit: IEE Regulations Part 2 B33, B40, B42, B53, B59, B70, B87–109, B125–127, B130, D3, D9, D29, K16, K22, K29.

Choice and construction of cables: IEE Regulations Part 2 B1–3, B5–24.

Exercises

1 Complete the following statements using one of the alternatives provided.

　a Where there is no risk of mechanical damage PVC sheathed cable

　　(i) may be installed without further protection,

　　(ii) should be protected by light-gauge conduit,

(iii) can be used in temperatures up to 90°C, (iv) can be replaced by unsheathed PVC for surface work

b The purpose of a skirt on a lampholder is to
(i) allow condensed moisture to run off, (ii) make fitting a shade easier, (iii) prevent touching the terminals when changing a lamp, (iv) reduce the risk of breaking a lamp

c Buckle clips are used to secure
(i) bell wire, (ii) VRI taped and braided cable, (iii) PVC insulated cable, (iv) PVC sheathed cable

d Conduit or trunking is generally used to protect
(i) unsheathed cables, (ii) mineral insulated cables, (iii) armoured cable, (iv) PVC sheathed cable

e When conduit is installed in damp conditions it is usually
(i) light gauge, (ii) galvanised, (iii) black enamelled, (iv) copper-plated

2 Select the correct answer to the following questions from alternatives provided.

a When bending conduit what is the radius of the smallest bend allowed?
(i) 12 cm, (ii) 4 times the conduit diameter, (iii) 2·5 times the conduit diameter, (iv) twice the total diameter of the cables contained

b What is the space factor to be applied to cables in conduit?
(i) 50%, (ii) 45%, (iii) 40%, (iv) 35%

c What should be fitted at the point where cables leave or enter conduits?
(i) an inspection elbow, (ii) a bush, (iii) a junction box, (iv) a circular terminal box

d What distribution system is best suited for open plan offices?
(i) wiring in conduit, (ii) overhead bus-bar trunking, (iii) wiring in skirting trunking, (iv) wiring in floor trunking

e What colour should be used for identifying electrical conduits?
(i) blue, (ii) orange, (iii) green, (iv) yellow

3 Why are specially designed accessories necessary for use with sheathed wiring?

4 Draw a neat cross-sectional sketch of a cord-grip lampholder suitable for use with an all insulated wiring installation.

5 Discuss the various points that could make it necessary to use conduit for certain installations.

6 Discuss the points which require attention when designing and constructing a conduit installation.

7 Describe three different types of trunking and the conditions for which each is particularly suited.

8 Summarise in simple terms the regulations which apply to trunking installation.

9 Describe three different situations in which mineral insulated cables would form the most suitable wiring system.

10 Compare conduit and mineral insulated cable installations with respect to mechanical strength, fire risk and ease of installation.

12
Bell and indicator circuits

Battery operation

The working of bells and buzzers has been described in Chapter 5 of this volume. Like all other electrical equipment they need a complete circuit if they are to operate. Most bells work with extra low voltage supplies. The simplest circuit consists of a battery, a bell push and a bell connected in series (Fig 12.1). A bell-push is

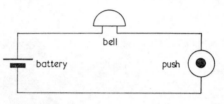

Fig 12.1 Battery operated bell circuit

simply a form of switch in which the contacts close when pressure is applied and re-open automatically when the pressure is removed. When the bell-push is pressed the circuit is completed and current flows through the coils of the bell causing it to ring. When the finger is taken off the push the circuit breaks and ringing ceases. This type of circuit can be used with either a single-stroke bell, a trembler bell, a buzzer or chimes.

Transformer operation

At first sight batteries seem to be a very useful source of power for bell circuits. In practice, however, the continued replacement or recharging of the cells used can be inconvenient. In order to avoid this inconvenience it is usual to supply bell circuits by way of a step-down transformer (Fig. 12.2). The transformer used will be

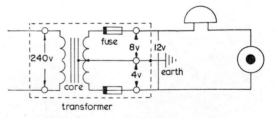

Fig 12.2 Transformer operated bell circuit

a single-phase one, stepping the supply voltage of 240 V down to 4 V, 8 V or 12 V, or as in the diagram giving a number of different secondary voltages. In order to comply with the IEE Regulations the outgoing or extra-low-voltage side of the transformer must be earthed and so must the metal core and the case if it is metal. The secondary side must be fused. Because the transformer illustrated has provision for connection at three points two of them must be fused so that there is no chance of an un-fused circuit being used.

Use of relays
The relays used in bell circuits are magnetically operated switches. Two different types of relay are used, one reopens automatically when its operating coil is de-energised and the other must be opened by hand or by means of a re-set circuit.

Reducing effect of volt drop
The first type of relay is used in circuits where the bell is a long way from the bell-push. In such cases a drop in voltage occurs between battery and bell. This can reduce the voltage at the bell to such a low value that the bell will not ring. To avoid this two batteries and a relay are used as illustrated in Fig. 12.3. It will be seen that

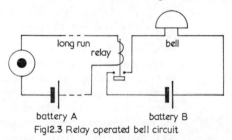

Fig12.3 Relay operated bell circuit

the circuit has two distinct parts. One part contains battery A, the bell-push and the relay-operating coil. The other part consists of battery B, the relay contacts and the bell. When the bell-push is pressed current flows from battery A through the long cable run and energises the relay coil. The magnetic field which results closes the relay contacts and completes the second part of the circuit. Current then flows from battery B, causing the bell to ring. With this arrangement battery A compensates for any voltage drop in the long run from the bell-push and battery B merely operates the bell.

Continuous ringing
It is sometimes necessary for a bell to carry on ringing after the bell-push has been released. This can be achieved by means of a special type of bell known as a continuous-ringing bell. Continuous ringing can also be effected by using the circuit described in the previous

paragraph. In this case the relay used must be a type with contacts that do not reopen as soon as the relay coil is de-energised. Continuous-ringing circuits may be used for fire or burglar alarms.

Closed-circuit alarm system

A simplified burglar-alarm system is illustrated in Fig. 12.4. It is called a closed-circuit alarm because the contacts of the alarm points are normally closed. Each alarm point is placed so that the opening of a protected

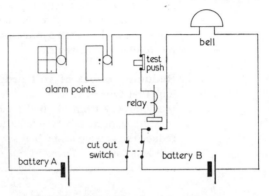

Fig 12.4 Closed circuit burglar alarm system

window or door allows its contacts to separate. All the alarm points are connected in series with battery A and the operating coil of the relay. When the circuit is in operation current from battery A flows through the alarm points and the relay coil energising it and holding the relay contacts open. If a protected door or window is opened a pair of alarm contacts are separated, breaking the circuit to the relay coil. The magnetic field in the relay coil collapses when the circuit is broken, and allows the relay contacts to close. This completes the circuit connected to battery B and the alarm bell rings. The cut-out switch allows the alarm to be put out of action when not required and the test button can be used to ensure that the circuit is working properly.

Bell and indicator circuits

Why indicators are necessary

When a circuit contains only one bell-push, if the bell rings it is obvious that the signal originates from that push. However, a circuit could be connected as in Fig. 12.5 with a number of parallel pushes each capable of completing the circuit and causing the bell to ring. Such a circuit might be used in a hospital or an office but all the pushes would need to be close together so that the person responding to the bell could easily find the origin of the signal. If the bell-pushes were spread over a number

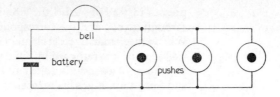

Fig12.5 Bell operated by parallel pushes

of rooms it would be necessary to use some means of showing which bell-push was causing the bell to ring.

Indicator boards are frequently used for this purpose. The most common form of indicator board is an enclosure with a dark glass front having a number of transparent 'windows', one for each bell-push. Each time the bell sounds a brightly coloured flag appears in the window corresponding to the bell-push used. In an alternative-type indicator lamps replace the flags, but these are less common. The flags in an indicator board are moved by the action of magnetically operated indicator elements. Some indicator elements are described in Chapter 5.

The basic indicator circuit

The pendulum type of indicator element and the mechanically reset element only require a simple circuit such as that in Fig. 12.6. The circuit shown

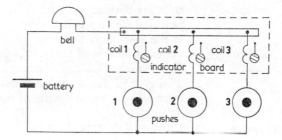

Fig12.6 Basic bell and indicator circuit

has three bell-pushes and each push is connected in series with the operating coil of an indicator element. This is called a *three-way bell indicator circuit*. The circuit can be extended by using a board with more elements and extra parallel connected pushes. Its operation is as follows. When bell-push 1 is pressed, current flows from the battery through push 1, coil 1 on the indicator board, through the common strap and the bell back to the battery. This causes the bell to ring and energises the coil causing its flag to move. The paths through the other coils remain open and therefore the other flags remain still. If either of the other pushes were pressed instead of push 1 the circuit would be completed through its coil and the corresponding flag would move. When a mechanically reset element is used the flag remains visible until it is returned by hand to its starting position. In the case of a pendulum element the flag will swing back and forth until it gradually loses momentum and stops.

Electrically reset indicator circuit

Figure 12.7 shows the circuit required for electrical replacement in a four-way bell and indicator system. It will be noted that there are eight coils—four of these

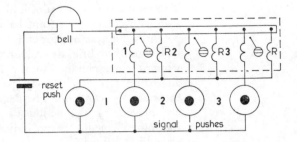

Fig12.7 Bell and indicator circuit with replacement coils

are for moving the flags and the other four are to return them to their original position when the signal has been noted. The returning process is called *resetting*. The operating coils are marked 1 to 4 while all the resetting coils are marked with an R. An extra push button is provided for 'resetting'. When the flags are positioned at the reset coils they are not visible. When the reset button is pushed all the reset coils are energised and any flags that are visible are returned to their starting place. This type of circuit can be expanded by the addition of extra bell-pushes and pairs of coils.

General notes

Circuits of the type described are often used in hospitals and hotels. Both the examples are shown as having batteries as their source of supply. In practice, however, in the interests of efficient operation it is more usual to supply such circuits from transformers.

Modified indicator circuits are used for fire- and burglar-alarm systems and also for various industrial applications such as high and low liquid level indicators.

Regulations for bell circuits

Some people have the idea that bell circuits are outside the scope of the IEE Regulations. This is not true. The only circuits not affected by the regulations are those which are not connected to power-distribution systems. Bell and call systems are dealt with by the British Standard Code of Practice (CP 327.401).

Bell and call systems: IEE Regulations Introduction (1) Part 2 A25, C32.

Segregation of fire-alarm circuits: IEE Regulations Part 2 B45–52.

Exercises

1 State which of the following statements is correct:
  a A single-stroke bell continues ringing as long as the circuit is closed. True/False
  b In any bell circuit a battery can be replaced by a suitable transformer. True/False
  c The IEE Regulations do not apply to bell circuits. True/False
  d The secondary side of a bell transformer must be fused. True/False
  e In a single-stroke bell circuit the supply is interrupted every time the striker reaches the gong. True/False

2 Insert the missing word or words in the following sentences using one of the alternatives suggested.
  a In a bell circuit with a long run of wire . . . . . may prevent efficient working
      (i) overheating, (ii) voltdrop, (iii) magnetic fields, (iv) high voltage
  b A relay-operated bell circuit requires at least . . . . . batteries
      (i) four, (ii) three, (iii) two, (iv) one
  c All the . . . . . must be connected in series in a closed-circuit burglar-alarm system
      (i) indicators, (ii) alarm contacts, (iii) bells
  d An indicator circuit may be needed when a bell is operated by . . . . . bell-push
      (i) a single, (ii) a concealed, (iii) more than one
  e An electrically reset indicator circuit needs . . . . . for every bell-push
      (i) two flags, (ii) two coils, (iii) one bell, (iv) one coil

3 Describe with the aid of neatly labelled diagrams the action of a relay-operated bell-circuit. What are the two different reasons for using such circuits?

4 Examine a bell indicator board, sketch one of the indicating elements and describe how it works.

5 Draw a neat diagram of a transformer-operated four-way bell and indicator circuit suitable for use with mechanically reset indicator elements.

6 What are the relative merits of pendulum and electrically reset indicator systems?

7 Draw the reset circuit of a three-way electrically reset indicator circuit and describe how it works.

8 How do requirements for an indicator system in a hospital vary from those in an hotel?

13

Elementary measurements and testing

Electrical indicating instruments measure current, potential difference and resistance by responding to an effect of a current or voltage. They show the value of the electrical quantity being measured by means of a pointer on a suitably graduated scale. Each instrument derives its name from the units of the electrical quantity it measures. Thus an ammeter measures current in amperes, milliamperes, or kilo-amperes; a voltmeter measures potential difference and e.m.f. in volts, millivolts or kilovolts, an ohmmeter measures resistance in ohms, kilohms or megohms, whilst a wattmeter measures power in watts or kilowatts, depending on their ranges and design.

All electrical indicating instruments have three main features—their modes of deflection, of control and of damping. The method of deflection determines the instrument's type and its suitability for measuring either the values of direct current and voltage or the effective values of alternating current and voltage.

Although various types of electrical instruments use various different effects of an electric current such as the heating chemical and magnetic effects to determine the magnitude of the current, a large number utilise the magnetic effect and are consequently known as electromagnetic instruments. They are classified as moving-coil permanent magnet, moving-iron and dynamometer types. The discussion of such instruments at this stage will be confined to the moving-coil permanent magnet and the moving-iron types.

Direct measurement of resistance can be made with an ohmmeter, but values of resistance may also be determined by several other methods—by means of the ammeter-voltmeter, the Wheatstone bridge principle, the substitution method, and the potentiometer. These are not discussed further at this stage of the course.

Moving-coil permanent magnet instruments

Fig 13.1 shows the construction of a moving-coil permanent magnet instrument. A soft iron core is held symmetrically by means of a brass bar between the soft iron pole pieces of a strong permanent horseshoe magnet made of cobalt steel. This core acts as a keeper, reducing the air-gap to a minimum and causing the main magnetic flux to cross the air-gap radially and uniformly. A rectangular aluminium former is pivoted to the iron core and carries a coil of fine, high-conductivity copper

Fig.13.1

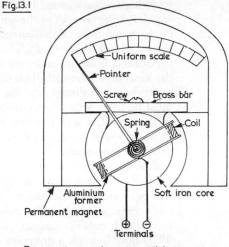

Permanent magnet moving-coil instrument

wire which is enamelled or cotton-covered for insulation. The ends of the coil are brought out to phosphor-bronze control springs—one left-hand wound and the other right-hand wound. These springs control the coil's deflection and also lead the current to and from the coil. The pointer attached to the coil is an aluminium tube which is flattened to a knife edge at the end that moves over the scale.

When current passes through the coil, the magnetic field created by the current interacts with the main magnetic field. The interaction of these two fields results in the coil deflecting and the pointer indicating on a uniform scale the value of the deflection (see Fig. 13.2).

When the pointer comes to rest, the deflecting torque due to the interaction of the magnetic fields equals the

Fig.13.2

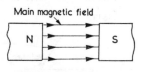

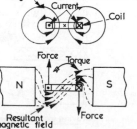

193

restoring torque exerted by the control springs, and the angular deflection (θ) of the coil is directly proportional to the magnitude of the current (I) that flows through the coil (in symbols, $\theta \propto I$). This relationship means that the scale is uniform, and may be calibrated in terms of the current in the coil. The direction of the deflection depends on the direction of the current. The instrument must be used with direct current, and the terminals are marked positive and negative to ensure that current is passed through the coil in the correct direction to deflect the pointer up-scale.

Efficient damping of the instrument enables the pointer and coil to reach a steady position quickly and without excessive oscillation. This is provided by the eddy currents that are produced in the aluminium former as it cuts the magnetic field (generator principle). They flow in such a direction as to oppose the motion producing them (Lenz's law) and thus exert on the coil a damping torque which depends on the rate of movement of the coil.

Moving-iron instruments

There are two main classes of moving-iron instruments —the attraction type and the repulsion type. In both a current passing through a coil of short axial length— that is, a solenoid—sets up a magnetic field which magnetises pieces of soft iron that are free to move. This action classifies the instruments as moving-iron instruments.

Attraction type of moving-iron instrument

Here the magnetic field created by the current flowing through the solenoid attracts a movable piece of soft iron into the solenoid (see Fig. 13.3). This produces a deflection which is independent of the direction of current flow, since the soft iron is always attracted towards the solenoid whichever way the current (I)

Fig.13.3

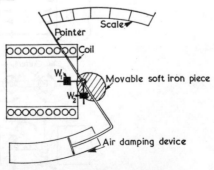

Moving iron instrument—attraction type

flows through it. The deflecting torque (T_d) produced in this way depends on the number of the ampere-turns on the solenoid. If the number of ampere-turns is kept low so that the soft iron does not saturate magnetically, deflecting torque (T_d) is proportional to the square of the current $(T_d \propto I^2)$.

The controlling or restoring torque (T_c) is provided by either gravity or spring control devices. Fig. 13.3 shows gravity control. With no current flowing, the pointer is set at zero on the scale by adjusting weights W_1 and W_2. Weight W_1 is a balance weight for the movement, and weight W_2 is the control weight. This weight exerts the controlling torque (T_c) when it is deflected from the vertical position shown in Fig. 13.3.

Gravity control has the disadvantage that the instrument has to be levelled before starting so that the weight W_2 is in the required vertical position. Also, with gravity control the controlling torque (T_c) is proportional to the sine of the deflection angle (θ) $(T_c \propto \text{sine } \theta)$. With spring control the controlling torque is proportional to the angle of deflection $(T_c \propto \theta)$.

In moving-iron instruments damping is always needed to prevent the pointer from oscillating. Air damping is always the method used since eddy current damping requires a permanent magnet and its field would affect that of the solenoid. This would result in inaccurate deflections. Damping is effected by enclosing a light piston of aluminium in a fixed chamber. The air in the chamber damps the oscillation of the moving system by retarding the movement of the piston.

When the deflected pointer is steady at a deflection angle, θ, the deflecting torque (T_d) equals the controlling torque (T_c).

The scale of a moving-iron instrument is an uneven scale. It may be made less so by appropriately shaping and positioning the soft iron.

Since the deflection torque is proportional to the square of the current $(T_d \propto I^2)$, the moving iron takes up a position that corresponds to the mean torque and the mean value of the square of the current. The instrument is therefore independent of current direction and is consequently suitable for measuring either the root mean square value of an alternating current or the value of a direct current.

Repulsion type of moving-iron instrument
Fig. 13.4 shows the repulsion type of moving-iron instrument. The current (I) to be measured passes through a solenoid and sets up a magnetic field. This field magnetises the two pieces of soft iron in the same direc-

tion so that the fixed iron repels the movable iron piece. In this way a deflecting torque (T_d) is produced which is dependent on the number of ampere-turns on the solenoid and proportional to the square of the current $(T_d \propto I^2)$. The pointer deflects and comes to rest when the deflecting torque equals the controlling torque (T_c) of the two phosphor-bronze springs. Damping again is effected by air damping.

Fig.13.4

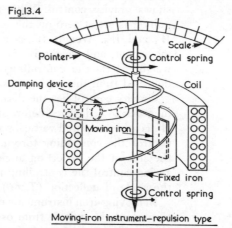

Moving–iron instrument–repulsion type

Both types of moving-iron instrument are suitable for use with alternating or direct currents. At the instant the direction of the magnetic field reverses with the alternating of the current, the polarities of the moving and the fixed soft-iron pieces change. For this reason the deflecting torque is always in the same direction.

Since the deflecting torque is proportional to the square of the current $(T_d \propto I^2)$ the scale is uneven, being cramped at the lower end and open at the upper end. This disadvantage can be overcome by suitably shaping and positioning the fixed iron.

The movement is enclosed in an iron case to prevent any stray magnetic fields from affecting the accuracy of the instrument. Spring control shown in Fig. 13.4 enables the instrument to be used in any position. For most applications, the repulsion type of moving-iron instrument has now replaced the attraction type.

Use of instruments

Use of ammeters and voltmeters
The uses of ammeters and voltmeters have been discussed briefly in Chapter 4 on pages 49 and 56 respectively.

Moving-coil ammeters and voltmeters
The moving-coil element, or galvanometer, is in effect a small d.c. motor whose armature moves over a limited

range—the armature's maximum rotation is obtained when only a small current of the order of 20 mA flows through the coil. The galvanometer is thus a direct-reading milliammeter of range 0 to 20 mA. If the resistance of the coil is 5 Ω, then the instrument is also a direct millivoltmeter of range 0 to 100 mV. It is a polarised instrument suitable for use in d.c. circuits, since the deflecting torque reverses when the direction of the current through the coil changes. A moving-coil galvanometer often has a centre-zero scale uniformly marked so that the magnitude of forward and reverse deflections may be observed.

A moving-coil instrument measures direct currents and voltages and the mean values of alternating current and voltages. Since the mean value of sinusoidal waves is zero the moving-coil instrument would also register zero, thus it is not used in a.c. circuits.

The range of a moving-coil galvanometer may be extended so that it can be used as an ammeter, by the addition of suitable parallel resistors called shunts; or, in order for it to be used as a voltmeter, by the addition of suitable series resistors to the instrument which are then called multipliers.

A measuring instrument must not affect the circuit in which it has been connected. An ammeter is connected in series with the part of the circuit through which the current flow is to be measured. Therefore an ammeter's resistance should be as low as possible in comparison with the resistance of the original circuit so that it adds as little as possible to the circuit. A voltmeter is connected in parallel with a part of a circuit to measure the potential difference or the voltage drop between two points in a circuit. For a circuit to be unaffected by the addition of a voltmeter the voltmeter's resistance should be very large in comparison with the resistance between the two points. Ideally a voltmeter's resistance should be infinitely large. If a low-resistance voltmeter were used, it would act as a short circuit between the two points in the circuit. This would also be the danger if any low-resistance instrument were connected in parallel with any part of a circuit. It is also the reason why ammeters must be connected in circuit with great care, to ensure that they are connected in series and not in parallel with the part of the circuit through which the current flow is to be measured.

Power loss (P) in an instrument must also be kept as low as possible. This power loss P in watts is given by:

$$P = I^2 R \text{ watts}$$

or

$$P = \frac{V^2}{R} \text{ watts}$$

where P = power loss in the instrument in watts
I = current through the instrument in amperes
V = potential difference across instrument in volts
R = resistance of the instrument in ohms.

When the instrument is used as an ammeter, it measures the current (I) flowing through it. Therefore for the power loss (I^2R) to be low, its resistance (R) must be small.

When the instrument is used as a voltmeter, it measures the potential difference (V) between its terminals. Consequently its resistance (R) must be a high value to keep its power loss (V^2/R) to a low value.

Moving-iron ammeters and voltmeters

In a.c. circuits a moving-iron instrument measures the root mean square value of the alternating current passing through its solenoid. When used as a voltmeter, this current is of the order of milliamperes and is proportional to the r.m.s. value of the alternating voltage that appears across its terminals. The instrument is in effect a milliammeter but its scale is graduated in volts.

The range of the instrument as a voltmeter may be extended by adding suitable series resistors or multipliers. However, it is not advisable to extend the range of a moving-iron ammeter with shunts because of the larger power losses in this type of instrument. Several ranges may be obtained by winding solenoids with different numbers of turns so that the product of the current and the number of turns for the various ranges produce full-scale deflection. It is therefore not used as an ammeter for multi-range d.c. measurements. However, for a.c. measurements its current range may be changed by means of a current transformer instead of a shunt.

A moving-iron instrument may be used in either a.c. or d.c. circuits.

Use of ohmmeters

An ohmmeter is an instrument which has a scale graduated in ohms or multiples or sub-multiples of ohms. When an unknown resistor is connected across its terminals the ohmmeter measures the resistance value of the unknown resistor directly in ohms.

A simple circuit of a battery-type ohmmeter is shown in Fig. 13.5.

To use the ohmmeter, terminals C and D are short-circuited and the rheostat R adjusted until full-scale deflection is obtained on the milliammeter mA. This corresponds to zero resistance external to the instrument. When terminals C and D are open-circuited, no current flows, since the external resistance between the terminals

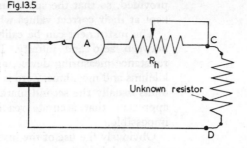

Fig.13.5

Unknown resistor

C and D is very large; that is, it approaches infinity (∞).

Thus the current scale can be graduated directly in ohms when the instrument acts as an ohmmeter. The range of the instrument can be designed to cater for high or for low values of resistance. When designed for low values it is referred to as a continuity meter, and when for high values as an insulation resistance tester. However, with this type of ohmmeter the p.d. across the resistance under test is usually small, since the cell e.m.f. is only about 1·5 V.

In many instances it is desirable to test the resistance at its normal working voltage, especially when conducting insulation-resistance tests on cables and apparatus.

The ohmmeter used for these insulation tests is the meg insulation tester (megger). It is shown in the circuit diagram in Fig. 13.6. The instrument has a hand-driven d.c. generator and a direct-reading moving-coil ohmmeter, in a case fitted with terminals marked line and earth, and sometimes with a guard terminal. The guard terminal and ring are attached to the case and the common terminal of the generator, to prevent stray currents from affecting the instrument reading. A clutch mechanism designed to slip at a predetermined speed is

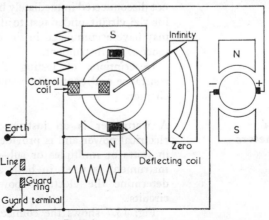

Fig.13.6

provided, so that the speed and the testing voltage are kept at their correct values when testing.

The instrument can be calibrated to read resistances between zero and infinity. It is essentially a high-resistance measuring device, and its scale is marked in kilohms and megohms. Zero is the first mark and 10 000 ohms usually the second mark of the scale, so one can appreciate that accurate reading of low resistance is impossible.

Obviously the use of the instrument is mainly limited to determining whether or not insulation in an electrical circuit is satisfactory.

The external resistance may vary from infinity for good insulation or an open circuit, to zero for poor insulation or a short circuit.

When an ohmmeter is used to measure the resistance of resistors, conductors or the insulation resistance of circuits the items under test must be isolated from all sources of supply except that provided by the ohmmeter. A useful precaution to observe before using any ohmmeter is to check that the instrument functions correctly. This is done by 'short circuiting' its terminals with a piece of wire and verifying that the instruments read zero when the instrument's supply is applied to the test circuit by turning its handle or by pressing the switch button. When the 'shorting' wire is removed and the instrument again operated the needle should then point to infinity.

To use a megger the leads attached to the line and earth terminals of the instrument are connected to the circuit under test which has previously been isolated. The handle of the instrument is then turned at a constant speed until the pointer is steady and shows the value of the insulation resistances of the circuit. The leads are not disconnected immediately but a short time is allowed for the circuit under test to discharge any charge that may have accumulated in the circuit through the resistance of the instrument.

Descriptions of testing circuits using insulation resistance and continuity testers are given later in this chapter.

Use of wattmeters

A wattmeter is an instrument used for measuring electrical power and is provided with a scale graduated in watts or multiples or sub-multiples of watts. This instrument is used in both a.c. and d.c. circuits to determine the electrical power consumed in those circuits.

Fig. 13.7 shows the connections of a wattmeter for measuring power supplied to a load.

Fig.13.7

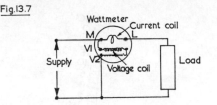

Note that the current coil of the wattmeter is connected in series with the load in a similar manner to an ammeter, whilst the voltage-coil circuit is connected in parallel with the load in a similar manner to a voltmeter.

Basic circuit-testing using insulation resistance and continuity testers

Although both types of ohmmeter – the insulation-resistance tester and the continuity tester—have been described in an earlier section of this chapter, their descriptions are briefly restated below in order to avoid any confusion between the two instruments.

Both of these instruments measure resistance: the insulation resistance tester is designed to measure large values of resistance, while the continuity tester measures small values. It will be remembered that the basic unit of resistance is the ohm (symbol Ω). Very large values of resistance are given in megohms (symbol $M\Omega$). One megohm is equal to one million ohms.

Insulation-resistance tester

This in its usual form consists of a hand-operated generator and a permanent magnet type of moving-coil measuring instrument. The movement of the instrument's indicating needle depends on the voltage generated when the handle of the generator is turned and the current which it causes to flow in the circuit under test. The voltage and current affect separate coils which are in opposition to one another, and the scale of the instrument can be marked off in terms of units of resistance with a maximum reading ∞ (infinity). See Fig. 13.8.

The operating voltage of these testers for use on the low and medium voltage range is 500 V. The current output from them is limited so that they cannot do any damage when used on low-resistance circuits.

Continuity tester

The continuity tester operates on the same principle as the insulation resistance tester. Because it is generally used on continuous conductors it is not expected to deal with large values of resistance. It will not therefore require a high voltage to circulate current, and accordingly, instead of using a hand generator, will be powered

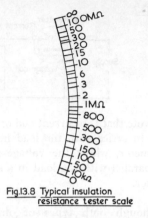

Fig.13.8 Typical insulation
resistance tester scale

by dry cells. The scale will be given in ohms with typical
maximum values of 100 Ω or 30 Ω. (Fig. 13.9). To

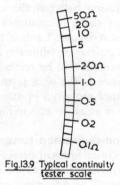

Fig.13.9 Typical continuity
tester scale

save the battery from continuous discharge a switch
is included in the circuit to be pressed when a reading
is required. More sophisticated types of continuity
tester have a switching arrangement to give a choice of
ranges of resistance.

**Insulation
testing**

In a healthy installation the only connections between
the line and neutral conductors will be where the loads
are situated: that is, at the lamps or other appliances.
Everywhere else there will be insulation between line
and neutral and of course between the currents carrying
conductors and earth. If therefore the lamps in an installation are removed and all appliances disconnected
there should be a very high resistance between lines and
neutral or earth. The purpose of an insulation test is to
check whether this is in fact the case.

Testing between conductors

The main switch controlling the circuit should be 'off'

or the supply not yet connected. All the auxiliary switches in the circuit must be 'on', the lamps removed and any appliances disconnected. The tester leads are held together and the handle turned slowly. When this is done the needle should move to the 'zero' end of the scale. The leads are then separated and the handle turned, when the needle should move to the infinity end of the scale. After thus checking that the tester is operating correctly the leads are connected to the line and neutral circuit terminals as shown in Fig. 13.10.

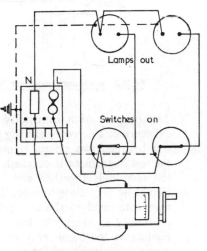

Fig.13.10 Insulation resistance between conductors

The tester handle is then turned steadily and a high value of resistance must be recorded if the insulation of the circuit under test is satisfactory. The precise values of reading which should be obtained will be discussed in the next volume of this series, but 1 M may be taken as a reasonable minimum value.

Resistance to earth

The circuit is prepared as in the previous test. The test instrument is checked and the leads connected as shown in Fig. 13.11. It will be noticed that in this case one lead is taken to a common line and neutral connection while the other is connected to earth. When a reading is taken under these conditions high values of resistance should be recorded as they were in the last test.

If both these tests show good results there is no danger of excessively high currents flowing when the supply is connected. They do not, however, eliminate the possibility of other faults such as open circuits.

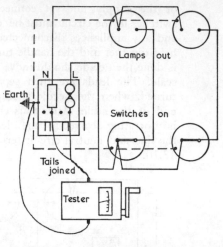

Fig.13.11 Insulation resistance from conductors to earth

The purpose of a continuity test is to establish the existence of a current-carrying path. Before a continuity test between two points is made, the effectiveness of the test instrument is checked. This is done by holding the test leads together and taking a reading, which should be very close to zero. The ends of the leads are then connected to the two points between which the continuity is being checked, and a reading taken. A reading at the 'zero' end of the scale will indicate a continuous circuit, whilst an open circuit will cause the needle to move to the high resistance end of the scale.

Continuity tests can also be carried out with insulation-resistance testers, which will show a zero reading when connected to a continuous path, or bell-test sets. A bell-test set is simply a unit consisting of bell and battery connected in series, with the leads required to complete the circuit brought out. When the leads are touched together or connected to opposite ends of a continuous electrical path, the bell rings.

Polarity testing

One of the most frequent applications of continuity testing is to ensure that the line conductor in a lighting installation is connected to the switch and not to the lampholder. The correct method of connection, or correct polarity as it is more often termed, allows lamps to be changed or minor adjustments to be made at the lampholder without risk, provided that the single-pole switch is turned off.

The test to determine whether this is the case is called a polarity test and is quite simple to carry out.

With the supply disconnected, one of the terminals of the single-pole switch controlling the circuit under test is earthed. The tester is then checked as previously described, and one lead connected to line and the other earthed at the distribution board controlling circuit (Fig. 13.12). The lamp or lamps in the circuit under test are removed and the single-pole switch turned on. A reading is then taken with the tester. If the reading is very low or zero then the circuit polarity is correct.

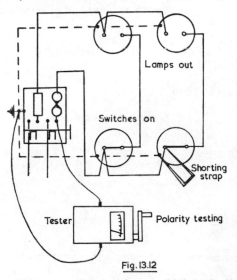

Fig. 13.12

If it is necessary to know which is the line terminal at the switch the reading is repeated with the switch off. If the reading remains low the terminal connected to the earth is the 'line': and if it becomes high the non-connected terminal is the 'line'. This test can also be made with an insulation-resistance tester or a bell set.

Exercises

1 Describe, with the aid of a sketch, the construction of a moving-coil instrument. Show, with the aid of a circuit diagram, how this instrument can be used to measure **a** voltage, **b** current. (W.J.E.C.)

2 Describe with sketches the construction of a moving-iron ammeter to measure up to 5 A. Show clearly how control and damping are obtained. (W.J.E.C.)

3 Describe with sketches the construction and operation of one of the following:
    a a moving-coil voltmeter
    b a moving-iron voltmeter.
State whether the instrument you have chosen may be used on a.c. or d.c. supplies, or both. (C.G.L.I.)

4 Show with the aid of a diagram how you would connect a voltmeter, an ammeter and a wattmeter to

measure the input to a single-phase a.c. load. What electrical quantities do each of these instruments measure?

5 Sketch an insulation resistance tester with which you are familiar and describe how it is used.

6 Draw a diagram of a circuit comprising two lights, each controlled by its own switch in which one wire is down to earth. The circuit is supplied from a distribution board. Describe how you would test the insulation of such a circuit and give the readings you would expect.

14
Electronics

Types of resistor

A resistor is a piece of apparatus designed to possess a certain resistance to the flow of an electric current through it. The value of the resistance that a resistor possesses is measured in ohms or multiples or submultiples of ohms.

Current in an electric circuit may be controlled by resistors and some of the types of resistors that are in common use are described below. Resistors are either made of a metallic resistance wire or a carbon composition or of a thin film of carbon. Thus there are two basic types of resistors—wire-wound resistors and carbon resistors. Either type may be made to have fixed or variable values.

As current flows through any type of resistor electrical energy is converted to heat energy by the resistance of the resistor. The rate at which the electrical energy is converted to heat energy is known as the electrical power consumed by the resistor. We have seen in Chapter 4 page 62 that the power in watts consumed by a resistor of R ohms resistance when carrying a current of I amperes is I^2R watts. As this electrical power is absorbed, the temperature of the resistor rises and the resistor behaves in effect as a heater. It is essential that a resistor is so designed that it is not damaged when it carries its current and absorbs power. The maximum current and power a resistor can carry without being itself damaged are known as its rated current and rated power respectively. It is also important that the heat produced by a resistor when carrying its rated current and power does not damage any other circuit component in its vicinity. Good circuit design and layout covers this aspect of a resistor's behaviour.

Thus when choosing a resistor for a particular application one should consider its resistance value, its rated current and rated power values. Also one should consider if there is any other component or device near the resistor that could be affected by the heat radiated from the resistor. Wire-wound resistors usually control much larger currents and powers than carbon resistors.

Wire-wound resistors
These resistors are made by winding a single layer of special resistance wire around a porcelain or ceramic tube and fitting metal terminals to the ends of the wire. The wire and tube are coated with powdered glass and

baked enamel so that the wire is protected and heat is conducted away from the wire. Such resistors are known as vitreous enamelled wire-wound resistors. Fixed tappings or sliders whose position can be changed may be fitted to these wire-wound resistors so that a fraction of the resistance value of the resistor can be selected if and when required.

For very accurate values of resistance over large temperature variations special precision resistors are made using manganin wire. These precision wire-wound resistors are expensive. They are used mainly as standard resistors and in test instruments.

Wire-wound variable resistors are made by winding resistance wire on a former which may be of a circular or linear shape with a rectangular section. A sliding contact is arranged to make contact with the wire in any position along an edge of the former. There are fixed connections made to the ends of the resistance wire and a connection made to the movable contact. The resistance of the wire between a fixed contact and the movable contact may be varied by altering the position of the movable contact relative to the fixed contact. See Fig. 14.1 for examples of wire-wound resistors.

Fig. 14.1 Wire wound resistors

Fixed wire wound resistor

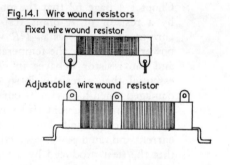

Adjustable wirewound resistor

Variable wirewound resistor

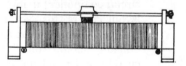

Another name used for a variable resistor is a rheostat. Power rating of a wire-wound resistor is relatively large compared with that of carbon resistors; power ratings of up to 50 watts, and above can be obtained.

Carbon resistors

Carbon resistors may be made from a rod of a carbon composition with wires fixed axially to the ends of the rod. Older types of carbon resistors were fitted with radial wire leads. The carbon rod is usually painted or coated with an enamel or a ceramic for protection. Another type of carbon resistor is made by depositing a coating or film of carbon on a ceramic tube with suitable connections made at each end of the tube to the carbon. The deposit of carbon in some cases is in the form of a spiral around the tube. A coating of baked enamel is again used to protect the carbon deposit and to conduct the heat away from the carbon to prevent the thin carbon deposit from overheating and burning out. In some forms of construction the carbon resistance element is further protected by being enclosed by a ceramic tube. See Fig. 14.2 for illustrations of some carbon resistors.

Fig.14.2 Carbon resistors

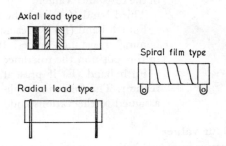

Variable carbon resistors are constructed by depositing a carbon layer on a circular insulated former. Connections are made to both ends of the carbon layer and a moving contact, that is operated by a rotating shaft, slides along the carbon track to vary the resistance between a fixed end terminal and the sliding contact.

This type of variable resistor is only capable of dealing with very small currents and powers. Power rating of carbon resistors is usually less than one watt, rarely does it exceed 2 W or 5 W. The power rating of carbon resistors must not be exceeded since they are liable to be irreparably damaged by excessive heating due to overloading.

Resistor colour code

For fixed resistors the colour code specified in BS 1852: 1967 entitled 'Marking codes for values and tolerances of resistors and capacitors' is used to indicate their resistance values and other characteristics such as tolerances and grades of the resistors. The method of marking is a series of circumferential colour bands on the resistor body (see Fig. 14.3).

With this coloured band method the significance of each band is:

First band (A) indicates the first significant figure of the resistance value.

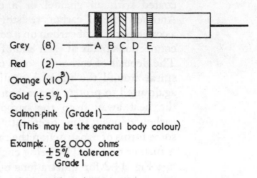

Fig.14.3 Coloured band marking

Grey (8) —— A B C D E

Red (2) ——

Orange (x10³) ——

Gold (±5%) ——

Salmon pink (Grade I) ——
(This may be the general body colour)

Example. 82 000 ohms
±5% tolerance
Grade I

Second band (B) indicates the second significant figure of the resistance value.

Third band (C) gives the multiplier.

Fourth band (D) if present gives the tolerance on the nominal resistance values. If no colour band appears in this position the tolerance is 20%.

Fifth band (E) if present denotes the grade of the resistor. The following table gives the numerical values assigned to the various code colours.

Table of colour values

Colour	1st figure (1st band)	2nd figure (2nd band)	Multiplying value (3rd band)	Tolerance % (4th band)	Grade (5th band)
Silver			10^{-2}	10	
Gold			10^{-1}	5	
Black			$1 (10^0)$		
Brown	1	1	$10 (10^1)$	1	
Red	2	2	10^2	2	
Orange	3	3	10^3		
Yellow	4	4	10^4		
Green	5	5	10^5		
Blue	6	6	10^6		
Violet	7	7	10^7		
Grey	8	8	10^8		
White	9	9	10^9		
Salmon pink					Grade 1
None				20	

A grade 1 carbon resistor with salmon pink 5th band or body colour is known as a 'high stability' type of resistor. These resistors are used in circuits where it is important that their resistance values are stable under all normal variations of temperature, humidity and time. An ordinary carbon resistor would be unsuitable in such cases since its resistance value can vary greatly with changes in temperature, humidity and time.

Older types of resistor, such as radial lead resistors, may employ a colour code known as the body-tip-spot method. This method and its slight variation (body-tip-band method) are illustrated in Fig. 14.4.

The colours used in either method have the same numerical values as those given in the above table.

Fig.14.4 'Body-tip-spot or band' code

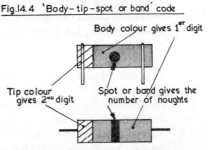

Body colour gives 1ST digit

Tip colour gives 2ND digit

Spot or band gives the number of noughts

Tolerance

It is difficult to make resistors with exactly the same resistance value as indicated by the colour code value or the nominal value of the resistor so a tolerance or variation on the nominal value is usually expected. For resistors with only three colour bands the tolerance allowed is 20%. This means that for resistors with no fourth colour band their resistance values may lie between 80% and 120% of its coded or nominal value. For example if a resistor has three colour bands brown, green and orange it has a nominal value of 15 000 ohms with a tolerance of 20%. Thus its actual resistance value could be anything between

$$12\ 000 \text{ ohms} \left(\frac{80}{100} \times 15\ 000\right) \text{ and}$$

$$18\ 000 \text{ ohms} \left(\frac{120}{100} \times 15\ 000\right)$$

A tolerance of 20% is acceptable for most applications but where greater accuracy is needed resistors with less tolerances of 10%, 5%, 2%, and 1% are available and must be used. These resistors would have a fourth colour band in the 'end to centre' band method and it would be coloured according to the table on page 210 silver, gold, red or brown respectively for the tolerances 10%, 5%, 2%, or 1%. In the older colour 'body-tip-spot or band' systems an extra spot or band appropriately coloured would give the tolerance value.

Thermionic emission

Suppose that a metal plate is enclosed in a glass envelope that has been evacuated of all gases and the metal plate or electrode is heated. Then as the plate's temperature rises some of the electrons of the electrode material acquire sufficient kinetic energy to be able to escape from its surface. The electrode that emits the electrons is known as the cathode. The rate at which electrons are emitted from the cathode depends on its temperature and the material of the cathode. This process of releasing electrically charged particles, namely the electrons, from the cathode by heat is called *thermionic emission*. Electrons released in this manner form a negatively charged cloud of electrons near the cathode. This cloud of electrons, called the space charge, tends to prevent more electrons escaping from the cathode since like charges repel each other.

The diode and its use as a rectifier

If a second metal plate or electrode is provided in the glass envelope a thermionic valve called a diode is obtained. A thermionic valve comprises a glass envelope containing two or more metal plates known as electrodes. Valves are named after the number of electrodes they contain by means of Greek prefixes. Thus valves with two, three, four or five electrodes are named diodes, triodes, tetrodes or pentodes respectively. In the hard or vacuum valve the glass envelope is evacuated of any gas so that the conduction of electricity must take place through a vacuum. If an inert gas such as argon or neon is introduced into the envelope the valve is referred to as a soft valve or gas-filled valve. In this instance conduction of electricity takes place through a gas. The discussion at this stage is limited to vacuum diodes.

When a second electrode called the anode is made positive with respect to the heated cathode, electrons are attracted from the space charge to the anode. This weakens the repelling effect of the space charge on electrons attempting to leave the cathode. Consequently more electrons are able to leave the cathode and travel to the anode. Eventually as the positive anode voltage is increased a stage will be reached when the number of electrons leaving the cathode equals the number entering the anode. When this happens the anode current ($I\hat{a}$) is said to be saturated. In this way conduction of electricity is achieved through a vacuum.

If the anode is made negative with respect to the cathode, electrons are repelled from the anode so that the movement of electrons from cathode to anode ceases. Therefore when the anode's potential is positive with respect to the cathode the vacuum diode conducts and when the anode's potential is negative with respect to the cathode the vacuum diode does not conduct. In this way the diode behaves as an on-off switch to current flow (similar to an on-off valve to water flow) when its anode potential is respectively positive and negative with respect to its cathode's potential. This behaviour enables the diode to function as a rectifier. The symbols used for various diodes are shown in Fig. 14.5.

Fig.14.5   Diode symbols

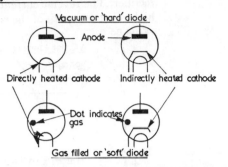

The cathode may be heated directly by a current passing through it or indirectly by an electrically separate tungsten-wire heater filament near the cathode. It is usually made from a nickel alloy which is coated with oxides of barium or strontium since they readily emit electrons when heated. Figs. 14.6 and 14.7 show circuit diagrams and current wave forms obtained using vacuum diodes for full-wave and half-wave rectification.

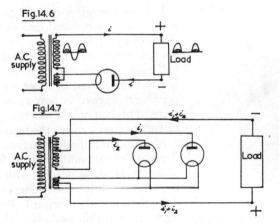

Fig.14.6

Fig.14.7

Alternating current (see Fig. 14.8a) may be rectified by devices known as rectifiers. A rectifier is a device that converts alternating current into a unidirectional current either by the suppression or the inversion of alternate half-waves of the alternating current. When alternate half-waves are suppressed the process is known as half-wave rectification (see Fig. 14.8b). If alternate half-waves of current are inverted the process is called full-wave rectification (see Fig. 14.8c). The ability of a rectifier to rectify alternating current depends on the fact that a rectifier allows current to flow easily through it in one direction, namely the forward direction, but prevents current flowing through it in the reverse direction. For half-wave rectification only the 'forward' current is allowed through the low resistance path of the rectifier. The reverse current is effectively blocked by the high resistance path of the rectifier.

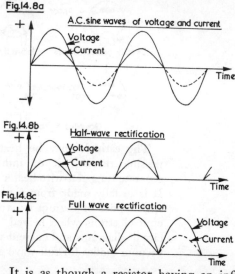

Fig.14.8a

A.C.sine waves of voltage and current

Voltage

Current

Time

Fig.14.8b

Half-wave rectification

Voltage

Current

Time

Fig.14.8c

Full wave rectification

Voltage

Current

Time

It is as though a resistor having an infinitely large resistance is introduced into the circuit for every negative half-cycle, for example, as if a switch automatically opened every negative half-cycle. Also for the positive half-cycle the rectifier behaves as a resistor having a low resistance, that is, it behaves as if the switch which previously opened is now closed for the positive half-cycle. This automatic switching is performed by the half-wave rectifier with the frequency of the a.c. supply.

A thermionic diode functions as a rectifier since its electron flow is unidirectional from cathode to anode while the anode is positive with respect to the cathode. It therefore only allows conventional current to flow through it in the forward direction from anode to cathode when the anode is positive with respect to the cathode.

Other types of rectifier are available, namely gas-filled diodes, mercury arc rectifiers, and semi-conductor rectifiers. Discussion on rectifiers at this stage is limited to thermionic vacuum diodes and semi-conductor rectifiers.

Semi-conductors in common use

The four types of rectifier using semi-conductor materials are:

    (i) the copper-oxide metal plate rectifier
    (ii) the selenium metal-plate rectifier

(iii) the germanium junction rectifier
and (iv) the silicon junction rectifier.

These semi-conductor materials when suitably treated are used to produce rectifier elements which offer a low-resistance path to forward current flow and a very high-resistance path to reverse current flow. Rectifiers using the semi-conductors copper oxide and selenium are known as metal-plate rectifiers. They are being superseded by the junction rectifiers that have the semi-conductor materials of germanium and silicon.

Copper-oxide rectifier

This type of rectifier element comprises a copper disc coated on one side with a very thin layer of cuprous oxide which is formed by heating at a temperature of 1000°C in a controlled atmosphere. The rectification takes place at the barrier layer between the copper and the copper oxide. Contact is made with the copper oxide usually by a soft-lead washer known as the counter electrode. Fig. 14.9 shows details of a copper-oxide rectifier element.

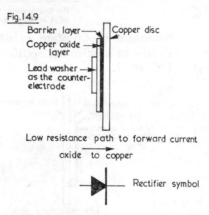

Fig.14.9

Barrier layer — Copper disc
Copper oxide layer
Lead washer — as the counter-electrode

Low resistance path to forward current
oxide to copper

Rectifier symbol

A copper-oxide metal plate usually has a number of rectifier elements connected in series, their number depending on the magnitude of the peak inverse voltage that the rectifier has to withstand. The size of the element and the number connected in parallel are decided by the magnitude of the current required to pass through the rectifier.

The rectifier elements are mounted on an insulated spindle with lead washers and cooling fins fitted between the elements. The lead washers make contact between the surfaces of the rectifier elements and the cooling fins. Connections to the external circuit are made from the cooling fins. There is a low resistance path for conventional current flow from oxide to copper and a

215

high resistance to flow of current in the reverse direction. By suitable arrangement of banks of rectifier elements half-wave or full-wave rectification can be achieved. Fig. 14.10 gives circuit diagrams using metal-plate rectifiers for half-wave and full-wave rectification.

Fig.14.10   Half-wave rectification

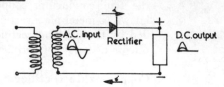

Full wave rectification

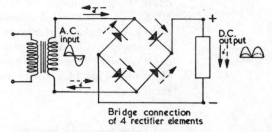

Bridge connection
of 4 rectifier elements

Selenium metal-plate rectifier

Details of a selenium rectifier element are shown in Fig. 14.11.

Fig.14.11

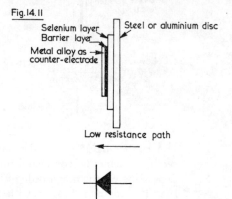

It comprises a steel or aluminium disc or plate which has a thin layer of selenium on one side. A metal alloy of lead, tin, bismuth and cadmium is sprayed on to the surface of the selenium to act as the counter electrode. The barrier layer is produced between the counter electrode and the selenium by special heat treatment. The direction of the low resistance path to forward current is from the basic plate metal to the counter electrode alloy for a selenium rectifier element. The

216

selenium metal-plate rectifiers are assembled and connected in similar ways to the copper oxide rectifier elements to achieve half-wave and full-wave rectification as shown in Fig. 14.10.

Germanium and silicon junction diodes

In their pure state the elements germanium and silicon are poor conductors of electricity. When certain impurities are added to them in controlled amounts they are said to be 'doped' with impurities. This 'doping' modifies their conductivities considerably so that they become known as semi-conductors. There are two kinds of impurities, namely 'donors' and 'acceptors'. A donor impurity such as arsenic, antimony or phosphorus gives or donates free electrons to the crystal of germanium or silicon to produce a semi-conductor crystal of the negative or n-type. An acceptor impurity such as indium, gallium, boron, or aluminium accepts electrons from the germanium or silicon, thereby producing a deficiency of electrons which is regarded as the production of positive holes. This type of crystal is then known as the positive or p-type. Although they are called negative and positive types, the semi-conductors are still electrically neutral but due to the presence of free electrons or mobile holes they are able to support conduction of electricity better than pure germanium and silicon.

If a junction is formed between the two types of germanium or the two types of silicon, a semi-conductor diode is obtained which is capable of functioning as a rectifier. These pn junction diodes allow conventional current to flow easily through them in the forward direction, that is, across the junction from 'p' to 'n' type when the applied voltage makes the p-type positive with respect to the n-type region. When the junction is reverse biased by the applied voltage making the p-type region negative to the n-type region, the junction diode prevents any appreciable current flow in the reverse direction from n to p regions. Thus the semi-conductor junction diode behaves in a similar manner to a thermionic diode. It is also able to rectify alternating currents since it offers a low resistance path to forward current flow when forward biased and a very high resistance path to reverse current flow when reverse biased.

The constructional details of germanium and silicon junction diodes are shown in Fig. 14.12 and Fig. 14.13 respectively.

Half-wave and full-wave rectification is achieved using junction diodes in circuits similar to those shown in Fig. 14.10 for metal-plate rectifiers. The graphical

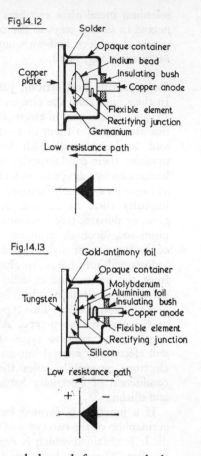

Fig.14.12

Solder
Opaque container
Indium bead
Insulating bush
Copper anode

Copper plate →

Flexible element
Rectifying junction
Germanium

Low resistance path

Fig.14.13

Gold-antimony foil
Opaque container
Molybdenum
Aluminium foil
Insulating bush
Copper anode

Tungsten

Flexible element
Rectifying junction
Silicon

Low resistance path

symbol used for a metal-plate rectifier and a semi-conductor diode is also shown in these diagrams. The arrow head in the symbol points in the direction of conventional current flow, that is, from anode to cathode in the thermionic diode and in the semi-conductor diode from p- to n-type material.

Limits of application, advantages and disadvantages of various rectifiers

The limits of application of various rectifiers depend on their operating voltage, operating temperature and the magnitude of current that the rectifier is capable of handling. Directly heated thermionic diodes have filament temperatures of 2000°C for a pure tungsten filament, 1600°C for a thoriated tungsten filament and 750°C for oxide-coated filaments. Their current ratings are approximately 5 mA, 60 mA and 100 mA respectively for each watt of power absorbed by the filament. The operating anode voltages of the tungsten filament are usually very high. Thoriated tungsten filament diodes are usually employed for values operating with anode voltages of 1000–5000 V whereas oxide-coated filament

218

diodes have much lower anode voltages and are generally used in smaller valves.

A diode with an indirectly heated cathode can use an a.c. supply for its heater and the heaters of several diodes can be supplied from the same a.c. source.

Metal-plate rectifiers, namely copper oxide and selenium, are used with current densities up to 1 mA/mm². They are cheap and robust. The copper oxide type is used mainly in low power circuits, for example, in instruments in telecommunications. Its operating temperature is around 55°C compared with 70°C to 90°C for the selenium rectifier. A copper oxide rectifier element is operated with a reverse voltage of 10 V compared with 30 V for a selenium rectifier. This means that fewer selenium rectifier elements are required to withstand a given peak inverse voltage. Selenium rectifiers are widely used in units of up to 10 kW rating.

The germanium and silicon junction diodes have current densities of between 1 A/mm² to 3 A/mm² which are 1000 to 3000 times more than the current densities allowed for metal-plate rectifiers. The operating temperature of a germanium junction diode is 35°C compared with 150°C to 200°C for a silicon diode. For power ratings greater than 10 kW where the direct voltage is less than 100 V a germanium junction diode is used. A silicon diode has a higher voltage drop and thus a lower efficiency than a germanium diode and is used when the direct voltage is greater than 100 V.

Exercises

1 What is a resistor? Name the various types of resistor that are in common use. Describe any one of these resistors.

2 If the colour markings on certain resistors are as follows what are their nominal resistance values?

Resistor	First band	Second band	Third band
A	brown	black	green
B	blue	grey	red
C	black	yellow	black

3 If the tolerances of the above resistors A, B and C are 5%, 10% and 20% respectively what will be the colour of the fourth band if present?

4 With the above tolerances what could be the maximum and minimum resistance values of resistors A, B and C?

5 What is 'thermionic emission'? Explain how this effect is utilised in the thermionic diode to rectify alternating currents.

6 Describe briefly with sketches a metal-plate rectifier. Show with the aid of diagrams how one or more of these rectifier elements are connected to provide **a** half-wave and **b** full-wave rectification.

7 Explain what is meant by the term 'semi-conductor'. Describe briefly a junction diode.

8 Show with the aid of diagrams the meaning of 'rectification' of an alternating current.

9 Describe one form of single-phase rectifier and explain its operation.